The Role of the Chimpanzee in Research

Symposium, Vienna, May 22–24, 1992

The Role of the Chimpanzee in Research

Volume Editors *G. Eder,* Vienna
E. Kaiser, Vienna
F.A. King, Atlanta, Ga.

42 figures and 12 tables, 1994

Basel · Freiburg · Paris · London · New York ·
New Delhi · Bangkok · Singapore · Tokyo · Sydney

Library of Congress Cataloging-in-Publication Data
Symposium on the Role of the Chimpanzee in Research (1992: Vienna, Austria)
The rôle of the chimpanzee in research/Symposium on the Role of the Chimpanzee in Research, Vienna, May 22–24, 1992; volume editors, G. Eder, E. Kaiser, F.A. King.
Includes bibliographical references and index. (alk. paper).
1. Chimpanzees as laboratory animals – Congresses. I. Eder, G. (Gerald), 1939–.
II. Kaiser, Erich, 1925–. III. King, Frederick A. IV. Title.
[DNLM: 1. Chimpanzee troglodytes – congresses. 2. Disease Models, Animal – congresses.
3. Animal Welfare – congresses. QY 60.P7 S989r 1994]
R853.A53S95 1992 619′.98–dc20
ISBN 3-8055-5850-3

Printed on acid-free paper
ISBN 3-8055-5850-3

Contents

II. Relationship between Chimpanzee and Human Development, Behavior and Reproduction

III. The Role of the Chimpanzee in Biomedical Research

IV. Conclusions

Preface

The opening of the Hans Popper Primate Center, constructed by Immuno AG, offered a splendid opportunity to discuss a series of questions with reference to the role of the chimpanzee in research.

It was a great pleasure that almost all of the invited outstanding scientists in this field all over the world accepted the invitation to give a lecture on their specialty. The symposium was organized in cooperation with the Yerkes Regional Primate Center of Emory University, Atlanta, and of the Department of Medical Chemistry of the University of Vienna. Yerkes Primate Research Center is the pioneer in chimpanzee research in the USA and indeed the world. Founded in 1930 by Robert M. Yerkes, it is dedicated to the behavior and reproduction of chimpanzees as well as a variety of scientific programs extending from molecular biology to models of human medical disorders and research related to mental health and social welfare. Dr. Frederick King, head of the primate center, dedicated his life to the exploration of the unknown fields of primatology and thus contributed significantly to our knowledge of primates. He supported the organization of this symposium, not only with his outstanding knowledge of chimpanzees but offered us the collaboration of his excellent staff as well, some of whom were among the lecturers.

The Department of Medical Chemistry, usually dealing with human biomedical problems, became interested in the chimpanzee years ago, when joint research studies with biomedical companies involving chimpanzees made evident the lack of complete data on the biochemistry of the chimpanzee's physical system. And it is exactly this information which is imperative for research involving these animals. Prof. Erich Kaiser, head of this institute, deepened his knowledge in this field and was happy to accept the cooperation in this symposium. 'Progressive improvement in public health has been one of the triumphs of mankind. In the twentieth century, various processes have jointly resulted in a remarkable increase in life expectancy at all ages, largely because of the dramatic declines in the incidence of and mortality from nutritional deficiencies and acute bacterial and viral infections. The understanding of the etiology of tuberculosis and poliomyelitis is based on animal experiments first performed by Koch and Landsteiner, demonstrating the provocation of both diseases by bacteria or viruses in animals. Today, as a result of all these efforts, both diseases that once accounted for a high death rate have nearly disappeared. However, medical doctors are today confronted with a series of new infectious diseases for which no effective treatment has yet been found. Successful prevention and treatment of these disorders require detailed information on the safety and efficacy of a new vaccine and the antiviral activity of a new drug, respectively. The majority of these data can only be provided by animal experiments. In order to obtain data applicable to man, they must be derived from animals closely related to the human species.'

The aim of this symposium was to treat the topic of the chimpanzee in research in a unique complex of scientific fields. Starting with the historic background of animal research, relating it to human experimentation, up to the latest knowledge and results in viral disease research, all fields were covered.

A pioneer in liver diseases and the 'father of hepatology', Hans Popper contributed throughout his life to our actual knowledge of the liver and its functions and diseases. He was the friend and teacher of many of the participants in this symposium; all who had the opportunity to work with him admired his tremendous energy in his work and his love for people. It was an honor for Immuno AG to dedicate the new primate center to his memory.

We want to take the opportunity to thank all speakers, chairmen and participants who made this symposium possible and who contributed to the success of it. The realization of the symposium might not have been possible without the enthusiasm and help of Dr. Benjamin Blood. Doctor of Veterinary Medicine and Master of Public Health, he served from 1975 to 1979 as executive director of the Interagency Primate Steering Committee of the National Institutes of Health in the USA and continued as consultant to the World Health Organization and the PanAmerican Health Organization. His knowledge of primates contributed considerably to the design of the Hans Popper Primate Center. His sudden death in January was mourned by colleagues and friends alike.

Prof. Friedrich Deinhardt, colleague and friend of many of the participants, died 3 weeks before the symposium. He started his career with his graduation summa cum laude and moved in 1954 from Hamburg to the USA where he served as researcher and professor, first in Philadelphia, then in

Chicago. From 1977 on, he was the director of the Max von Pettenkofer Institute at the Ludwig Maximilians University in Munich. He had high expectations for our symposium, and his scientific work was crucial to the research in infectious diseases.

The symposium on 'The Role of the Chimpanzee in Research', retrospectively seen, showed the ongoing need for these animals to ensure the well-being of humans by the development of medications and vaccines in the fight against life-threatening diseases. Their wise and responsible use in biomedical research for the benefit of all of mankind is irreplaceable.

G. Eder
E. Kaiser
F.A. King

The New Primate Center of Immuno AG at Orth/Donau: An Introduction to the Design of the Building

Gerald Eder

Hepatitis research has always played a central role in the policy of Immuno AG. In the early seventies, experimental studies were carried out on chimpanzees in the form of contract research in the USA and met with only limited success. Thus the establishing of the Company's own primate unit focussing on hepatitis B, hepatitis non-A/non-B research with emphasis on transmission and inactivation studies as well as safety and efficacy studies of hepatitis vaccines was inevitable. The basis for such a chimpanzee unit was established in the second half of the seventies by importing animals first from Lemsip, then from Africa, Belgium and TNO; some were purchased in Austria. An evaluation of the development of the body weight of our chimpanzees at the end of 1986 showed a need for larger cages. An expansion within the already existing building was impossible. Our research projects in the field of hepatitis and in the development of AIDS vaccines called for a new biocontainment facility that had to meet the following requirements: We were to construct a building of animal biosafety levels 3–4 for chimpanzees kept in individual housings [1]. We had to consider that chimpanzees might be infected with hepatitis B, C or HIV and consequently remain infectious and have to be kept in the building for several years. We had to bear in mind that almost all infectious diseases are transmissible from chimpanzee to chimpanzee, from chimpanzee to man and vice versa. We had to comply with all regulations concerning animal rights issued by the US Department of Agriculture, National Institutes of Health, American Association for Accreditation of Laboratory Animal Care, European Community and Austrian authorities, and with all regulations concerning security and health protection of employees (e.g. national statutory protection of employees, industrial safety regulations, Institutional Biosafety Committee

Regulations) [2, 3]. We had to use protective suits with external air supply in the sewage treatment area and in the area holding HIV-infected chimpanzees. Hypochlorite as decontamination solution had to be used for these suits and consequently all environmental regulations had to be observed. Visitors were to be allowed to see animals and treatment procedures without direct access to animal quarters. All technical facilities were to be strictly separated from the animal rooms as such. Thus, the first floor was designated for all supply facilities including water, air and electricity, where they could be accommodated in a noninfectious area and would be easy to service. The ground floor was chosen for animal housings, treatment rooms, administration, and kitchen as well as the preparation of liquid waste. We decided to construct 14 animal rooms with 4 individual housings each and a 2-way corridor system. All 4 individual housings had to be connectable and adaptable in size, as well as equipped with a movable back wall, side wall windows to see the neighbor chimpanzee and ceiling windows to see the skylight and plants. A prerequisite for the design was to exclude any body contact with the neighbor but to enable hearing and smelling as well as body contact with the animal keeper.

As a result, we built a 1:1 model of this new type of animal housing in an empty storehouse in 1988. Based on this model, the animal rooms were finally constructed. One animal room consists of 4 individual animal housings with concrete walls, movable back walls (maximum speed 5 cm/s = 2 inch/s) and front gratings. The concrete walls are covered in green (nontoxic coloring pigment bound in concrete), ceilings are brown. These individual animal housings may be connected by removing the glass panels of the side walls. Each individual housing contains ropes and a hammock. Videos and TV games may be watched through the window in the back wall. Front doors and back walls are motor-driven [4, 5]. Windows are made of laminated tempered glass with a thickness of 40 mm = 1.57 inch. The base of the animal housing is 0.7 m (= 27.5 inch) above the floor. This area is cleaned by a motor-driven machine with high-pressure water jets 3 times a day. Space allocated to the base is approximately 5 m^2 (= 53.82 square ft.), height 2.30 m (= 7.55 ft.). High-efficiency particulate air filters are used for supplied and exhausted air, guaranteeing a negative pressure relative to the surroundings to avoid pollution of the environment. A tropical climate with a temperature of 28 °C (= 82 °F) and a humidity of approximately 70–80% is simulated in each animal room. The design of the rest of the building was based on these animal rooms. Access to the rooms is provided by a supply corridor with 7 rooms to the left and 7 to the right side. In the so-called infectious area it was necessary to install a treatment center consisting of 2 preparation rooms, 2 treatment rooms and 1 room for the sterilization of instruments and surgical disinfection of the hands. In the old unit we had primarily used disposable materials for e.g. masks, caps

Fig. 1. The new primate center, with the castle of Orth in the background.

or coats. This led to an enormous amount of firm waste that had to be autoclaved and constituted a big problem with regard to waste disposal. Therefore we decided to do without disposable materials in the new building and designed a new sewage treatment concept. The entire waste from the animal rooms is cut into small particles with a diameter of 2 mm maximum by means of a slicing machine and is collected together with the liquid waste in a tank with a capacity of 8,000 liters. In the so-called kill tanks with a capacity of 4,000 liters, waste is sterilized (120 °C = 248 °F/30 min). Thus, the amount of firm waste that would fill 800 metal containers with a volume of 200 liters each (= 160,000 liters) and had to be incinerated before can now be avoided.

After approximately 60 planning meetings and the approval by building and industrial authorities on April 10, 1989, the construction of the building was started. The 4-storied building with sewage disposal in the basement, animal rooms, administration and treatment rooms on the ground floor, and technical facilities on the first and second floors, comprises 25,131 m^3 of enclosed area, with a length of approximately 80 m and a width of 28 m (fig. 1). The construction of the building consumed 4,000 m^3 of concrete and 168 tons of steel. The animal housings were made of prefabricated concrete parts, altogether 112 side walls and 56 rear walls, with 31 tons of concrete-reinforcing steel. A food kitchen is accomodated in the noninfectious section, where a yearly amount of 80 tons of fruits and vegetables is being processed.

Administration on site is reduced to a minimum, since the entire monitoring of the building can be controlled by one operator only.

References

1 Biosafety in Microbiological and Biomedical Laboratories. US Department of Health and Human Services, Public Health Service, Center for Disease Control and National Institutes of Health, March 1984, HHS Publ No (CDC) 84-8395, Washington, US Government Printing Office, 1984.

2 Killander J, Chippaux A, Eder G, Elliot B, Gürtler L, Wilde C: Protection of Laboratory Workers from Infectious Agents Transmitted by Blood and Other Biological Material. European Committee for Clinical Laboratory Standards, March 1990.

3 Miller BM, Gröschel DHM, Richardson JH, Vesley D, Songer JR, Housewright RD, Barkley WE: Laboratory Safety: Principle and Practices. Washington, American Society for Microbiology, 1986.

4 Eder G, Womastek K, Reiländer J, Simonich W: Cage Structure for Restraining Primates. Request for grant of a European Patent pending, Application No 89890098.0, April 7, 1989. European Patent Office, Patent No 0 391 016.

5 Eder G, Womastek K, Reiländer J, Simonich W: Cage Structure for Restraining Primates. United States Patent No 5,056,465. Date of Patent: Oct 15, 1991.

Gerald Eder, MD, Hans Popper Primate Center, Head, Clinical Research – Gastroenterology, Immuno AG, Uferstrasse 15, A–2304 Orth/Donau (Austria)

A Tribute to Hans Popper

Heribert Thaler

Only death opens the gate to glory. Our friend Hans Popper was an exception to this rule, he was already a legend during his lifetime.

Due to the rapid developments in the field of medicine in our century as well as the breadth of highly qualified scientific research, the individual is confronted with more and more difficulties in attaining a dominating position. If a person is to succeed, he needs favorable circumstances as well as outstanding personal capabilities. Hans Popper had both.

Hans was born on November 24th, 1903, into a wealthy family of physicians in Vienna. As a child he experienced the fading glory of imperial Vienna. Perhaps it was the impending decline that caused a unique bloom of art and intellectual life in Vienna. Representative of this era were Austrian, Jewish-cosmopolitan intellectuals whose work continued until after the lost war and was finally crushed by Nazi boots. Many of the fundamental intellectual ideas of our century were conceived in Vienna's coffee houses, studios and laboratories.

Hans would not have been Hans if he had not profited from this intellectual atmosphere. Under the circumspect guidance of his father he enjoyed a profound humanistic education, rounded out by academic training in biochemistry, pathologic anatomy and internal medicine.

In 1938 all preparations were completed for Hans's habilitation (a prerequisite in all German-speaking countries for an advanced university career). Only the signature of the President of the Federal Republic was missing when Hitler marched into Austria. The bottom fell out of Hans's world but not only his alone.

The next days clearly showed that Hans must have had a magnificent guardian angel. Several months previously he had been offered a position at Cook County Hospital in Chicago. However, he had delayed his acceptance, because he wanted to wait for his appointment as 'Universitätsdozent' (similar to the American Associate Professor). Now he only had to send a telegram to Chicago, and on the day when the Gestapo was to arrest him which would have meant his death sentence for sure, he took the last flight to Prague and thus to freedom. From now on he rapidly climbed the ladder to success: Director of Pathology at Cook County Hospital, founder of the Hektoen Institute for Medical Research in Chicago, successor to Klemperer as Pathologist-in-Chief at Mount Sinai Hospital in New York, founder and First Dean for Academic Affairs of the Mount Sinai School of Medicine, Gustav L. Levy Distinguished Service Professor for lifetime.

It is pointless to speculate what would have become of Hans if Hitler had remained a painter of postcards and Hans could have stayed in Vienna. Certainly he would have turned into a renowned scientist here as well, but to become the great Dr. Popper who was known and honored by every physician from Taipeh to Buenos Aires, it took the USA with its then still unlimited opportunities. But thousands of other physicians fled Hitler's tyranny and went to America without attaining what Hans achieved so easily. How could he have turned into one of the most famous and renowned physicians of the world, be more or less the founding father of hepatology and at the same time the motor of its rapid development? This can only be explained by someone who knew Hans Popper as a person.

He was one of the most hardworking people I ever met and had an enviable constitution. As a rule he worked 15 h a day, and still it is hardly conceivable how he could have managed to fit all his varied tasks, administration, lectures, dissecting room, histological routine and scientific work, lecture tours and study of literature into his tight schedule and still not leave any letter unanswered.

Hans detested holidays, but he was always fond of company. He was a lover of 'Pflümli' (plum brandy) and good wine, and congresses frequently lasted long into the wee hours. Nevertheless he was the first one to take his seat in the lecture hall the next morning, as if nothing had happened, while those of us who were much younger sneaked in more dead than alive during the course of the morning. Hans had a stupendous memory, and until his death he literally retained everything. He was ready to learn from everyone and everywhere and possessed a unique reasoning power, always recognized the essential things and was thus capable of reconstructing comprehensive pictures from bits and pieces that he picked up here and there. When he was wrong, he was already ready to admit his error, but this was something that was not required very frequently. Within our small group of liver experts, nicknamed 'the gnomes of Zurich' by Sheila Sherlock, and whose driving force was Hans for 17 years, we had the chance to admire his amazing capabilities every year.

As already mentioned, Hans enjoyed not only worldwide esteem but also worldwide love that he owed to other personality traits such as friendliness, kindness and the resulting readiness to help others. He patiently listened to minor colleagues from developing countries, readily gave information and helped whenever it was within his power.

You cannot blame emigrants for the lifelong bitterness they feel about how they were once treated in their native country. But Hans was different, he could forget and forgive. After our first meeting in 1953 in Chicago we met several times a year at numerous congresses and conferences, we travelled together, and he became a faithful, fatherly friend of mine, to whom I owe a great deal. Maybe the reason for our close relationshp was that I came from Vienna. Although he always underlined that he owed the USA and its people a great deal, a sentimental love bound him to his hometown. Moreover, I worked in the same hospital and even the same ward that used to be under his management. In addition, we had the same passion for hepatology, and liver histology in particular. Thus, whenever he introduced me to somebody, he always added 'my nephew'.

Hans Popper's great interest in primates was awakened in 1974 by the animal experiments carried out by Purcell and Gerin, and he soon became the leading expert in hepatitis histology and oncology in the liver of primates. We owe him numerous papers of fundamental importance in this field. Thus it was only a logical step that Immuno, one of our few companies of worldwide reputation, crowned the close friendship with Hans Popper by dedicating this primate center to him. Hans had chosen Morgagnis's words that decorate the lecture hall of the Institute of Pathology and Anatomy in Vienna and also his study in New York as his motto:

> Hic locus est ubi mors gaudet succurrere vitae.
> (This is the place where death gladly relieves life).

Slightly altering this sentence one could say about this center:

> Hic locus est ubi animal gaudet succurrere vitae humanae.
> (This is the place where the animal gladly relieves human life).

A tribute to Hans would be incomplete if it did not include the one person who played an essential role in his life, his wife Lina. Hans became acquainted with the charming Viennese lady Lina Billig in Chicago and fell in love with her. She not only gave him two able sons, she was also his ideal complement. With her quiet and clever nature she perfectly understood how to keep him on a long leash that was only tugged when impatience and stress were too much. For Hans would have preferably given 2–3 lectures every day and if possible on different continents! Lina was his best adviser and someone he was willing to obey – and this really means something.

Hans died on May 6th, 1988, half a year after his 84th birthday. He remained intellectually alert well into his advanced years, and it was literally death that took the pen out of his hand. More than 800 scientific papers and 28 books were the fruits of his scientific life. The USA and the world have thanked him with all honors imaginable already during his lifetime, among them 14 honorary doctorates.

To conclude and summarize, I find the following verse by Goethe from *Faust* quite apt:

Wie sich Verdienst und Glück verketten,
Das sehen Toren niemals ein,
Wenn sie den Stein der Weisen hätten,
Der Weise mangelte dem Stein.

Or in its beautiful translation by Philip Wayne:

How Merit comes to be with Fortune twined
Is to these fools undreamed-of and unknown:
Give them the Stone of Wisdom, and you'd find
Philosophy gone – and what was left, the stone.

Heribert Thaler, MD, Professor of Medicine, Sebastianplatz 7/5, A-1030 Vienna (Austria)

Hans Popper and Mount Sinai

Fenton Schaffner

Hans Popper had three simultaneous professional careers: first as a practicing physician, being both a clinician and a pathologist, secondly as a research scientist, combining many different disciplines of biology, and thirdly as an educator. This last was made in the classroom and in the laboratory and on a national and international level as the founder of AASLD and IASL. For us in New York, the most important was his role at Mount Sinai where he was the driving force for the conception, gestation, delivery and early development of our medical school. Hans Popper arrived at the Mount Sinai Hospital in 1957 as the pathologist-in-chief. By 1959 he convinced his colleagues that the future of Mount Sinai in world medicine could only be guaranteed by creating a medical school; the fact of starting a hospital-based school had rarely been accomplished in the USA. Within 2 years, Hans convinced the board of trustees, the people who were going to be largely responsible for getting the necessary funding. By 1963 the State of New York granted Mount Sinai the charter to start the school, and in 1965 construction began. Hans was responsible for making the affiliation with the City University of New York so that the Mount Sinai School of Medicine became one of the 13 colleges of the City University. Hans was the interim dean, and when the first students came in 1968, he became the dean for academic affairs. The first students graduated in 1970, and the first full 4-year class in 1972, at which time Hans Popper was dean and president. He held this position for 2 years when he recruited Tom Chalmers from the National Institutes of Health to replace him. The Annenberg Building, our tower of learning, was dedicated in 1974 by the then Vice President, Gerald Ford. In 1977 Hans was made the first Gustave L. Levy Distinguished Service Professor. He never retired but continued to work until shortly before his death. His last contribution appeared in the ninth volume of *Progress in Liver Diseases*, a series we began in 1961 when we were both new to Mount Sinai. Hans Popper was responsible for the growth of hepatology from a minor branch of gastroenterology to become one of the exciting areas of progress in modern medicine by integrating clinical medicine and the basic sciences to form a new discipline. The world will remember him for this contribution. Hard work, however, was not always the rule, and relaxation was necessary, even if only for brief periods, especially when it came to contemplating the beauties of nature near his beloved Vienna.

Fenton Schaffner, MD, Emeritus, George Baehr Professor of Medicine,
Mount Sinai School of Medicine, 1, Gustave L. Levy Place, New York, NY 10029 (USA)

Eder G, Kaiser E, King FA (eds): The Role of the Chimpanzee in Research.
Symp, Vienna 1992. Basel, Karger, 1994, pp 1–6

The Rights of Human Research Subjects and the Necessity of Conducting Animal Research as Illuminated by the Nuremberg Code and the Declaration of Helsinki

Charles R. McCarthy

Office for Protection from Research Risks, National Institutes of Health, Bethesda, Md., USA

In October 1946, following World War II, 23 German scientists were placed on trial for war crimes committed in World War II by the Nuremberg Military Tribunal (USA vs. Karl Brandt et al.) [1]. They were charged with committing crimes against humanity relating to experiments in which they subjected human beings to trials which included: cold water survival; decompression; transplantation; sterilization techniques, and techniques of massive killing of persons judged to suffer from chronic mental illness. Seven of the accused were sentenced to death, 7 were acquitted, at least one, Prof. H. Eppinger, committed suicide; Dr. S. Rascher died before the trial was completed, and the infamous Dr. J. Mengele escaped [2].

The Tribunal not only brought the scientists before the bar of justice on traditional charges of murder, assault and battery, but in its final deliberations the Tribunal issued a code for the conduct of medical research intended to set inviolable standards for research involving humans.

As a part of the rationale for its decisions concerning the guilt or innocence of the accused, the Tribunal promulgated a document known as the Nuremberg Code. The Code included 10 principles setting forth conditions for voluntary participation in research by competent, normal, adult human subjects. The Nuremberg Code combined traditional concerns for human beings – concerns that had been a part of western medical tradition at least from the time of Hippocrates – with more contemporary philosophical and western common law legal traditions that placed strong emphasis on the principle of respect for persons.

In the case of research involving competent adult human beings, the Tribunal asserted that respect for persons included acknowledgement of the radical autonomy of each person. Respect for personal autonomy was to be concretely expressed, according to the Code, by providing potential research subjects with the opportunity to make an informed choice to be involved or not to be involved in research. This choice is commonly referred to as the process of obtaining informed consent. Although far from new to the practice of medicine and to medical research ethics, the emphasis that the Nuremberg Code assigned to informed consent by making it the very first article and by stating that 'the voluntary consent of the human subject is *absolutely* essential' constituted a significant departure from the traditions of the past which had tended to emphasize beneficence (maximizing good) and nonmaleficence (minimizing harm) as primary and transcendent principles.

The worldwide publicity afforded the Nuremberg Code had a profound influence in many parts of the world. Among those influenced by it were the authors of the Charter of Human Rights of the newly formed United Nations. Many of the principles of Nuremberg were quickly endorsed by all of the signatory nations of the Charter of the United Nations. A few short years after World War II, the principles of Nuremberg – in part through the agency of the United Nations – provided a framework for worldwide recognition of the rights of individuals.

The Nuremberg Code did not directly address many contemporary questions of research ethics. The Code dealt exclusively with issues related to persons who have come to be known as adult 'normal volunteers', that is persons who were not expected to benefit in any direct way from their participation as subjects of biomedical research.

The Code did not address questions concerning how best to apply its principles to research intended to be beneficial to the subjects. It did not address research involving human fetuses, children, mentally disabled or mentally handicapped persons, persons suffering from dementia, women of childbearing potential and pregnant women or prisoners. Part of the impact of the Nuremberg Code can be attributed to the fact that it singled out the clearest case and made unequivocal statements pertaining to the rights and welfare of subjects in that case. Its message was unequivocal and poignant.

Perhaps because of the stark simplicity of the Code that did not attempt to address more complex and less straightforward cases of the rights of research subjects, the Nuremberg Code has had a profound effect. It strongly influenced the formulation of the United Nations' Declaration of Human Rights. Consequently the Nuremberg Code became, in effect, the first truly international code of research ethics, because its principles were endorsed throughout the world and because the views set forth in the Code were endorsed by all of the original signatory nations of the Charter of the United Nations. The Nuremberg Code enunciated principles that could be defended by at least two traditions of bioethics, the deontological and the utilitarian. Consequently it did not endorse a specific ethical school or tradition but enunciated rules that could be defended by several highly respected ethical traditions.

The first principle of the Nuremberg Code states: 'The voluntary consent of the human subject is absolutely essential.' Although this principle, enshrined in American Common Law [3], can be defended by utilitarian theory calling for the greatest good of the greatest number, it is more likely that the principle derived from a deontological tradition of ethics emanating from Emmanuel Kant [4]. According to Kant, it is a moral imperative admitting of no exception that human beings are to be used only as *ends*, never as *means*. In practice that means that it is always immoral to use competent adults as subjects of research studies unless the subjects are adequately informed and freely consent to participate.

Eight of the other 9 principles contained in the Nuremberg Code appear to derive from the Hippocratic tradition that requires each physician (or research investigator) to help or at least to do no harm – the principles of beneficence and nonmaleficence. These principles are often expressed in a utilitarian context that requires the physician to seek the greatest good, all things considered, for his or her patient [5]. The Nuremberg Code (principle 2) modifies the Hippocratic tradition in the research context by allowing (providing that subjects provide informed consent) research to be justified by an appeal to the good of society (rather than the individual patient). Principle 9 returns to a deontological framework by insisting that the research subject must *always* be free to withdraw from participation in a research study and thus to bring his or her participation in the experiment to an end irrespective of the value of the research and the effect on the research that his or her withdrawal would have.

Of particular interest to this conference dealing with nonhuman primates is principle 3 of the Nuremberg Code:

> The experiment should be so designed and based on the results of animal experimentation and a knowledge of the natural history of the disease or other problem under study that the anticipated results will justify the performance of the experiment.

No exceptions or limitations are placed on the principle that experimentation involving humans is to be justified by the experimentation involving animals. The Code seems to anticipate at least two different uses of animals that ought to precede involvement of human subjects in research.

First, the Code states that if a human person is to be subjected to an experiment, the experiment ought first to utilize animals as subjects in order to anticipate, so far as possible, the effect the experiment is likely to have on human beings. Data developed by using animals as research subjects will provide a basis for deciding whether the risks to subjects associated with the research are reasonable in the light of expected benefits.

Second, by including reference to studying 'the natural history of the disease or other problem' the Code seems to be calling for the use or development of suitable animal models to be studied as human surrogates to help the researcher to understand, so far as possible, the causes and natural history and to design methods for diagnosis, prevention, treatment or cure of the disease or condition.

The Nuremberg Code not only permitted animals to be used in a way that it would be unethical to use a human person, but the Code made it imperative that animals be used, wherever possible, as surrogates for human research subjects and as models for human diseases and conditions. Furthermore, animal subjects were to be used prior to any human involvement in the research.

There is a stark contrast between allowing or permitting the involvement of human beings in research, provided that they have been adequately informed and have freely consented to participate and requiring the involvement of animals as subjects prior to permitting human persons to be research subjects. It seems reasonable to infer that the Tribunal in no way recognized any right in animals not to be used as research subjects. Furthermore, it should be noted that the Tribunal did not restrict or condemn the use, for research purposes, of any species of animal, including the chimpanzee if such use were appropriate or necessary, to justify involving humans in research.

The statement of the tribunal relative to use of animals as research subjects prior to involvement of human research subjects is unequivocal. Failure to carry out appropriate preliminary research in animals was one of the many serious flaws in the research conducted by the convicted scientists under the Nazi regime.

If the interpretation of Nuremberg provided above is in error, it would surely have been corrected by the Declaration of Helsinki which was produced by the World Medical Association in 1964.

The Declaration of Helsinki – adopted by the 18th World Medical Assembly (Helsinki, June 1964) and amended by the 29th World Medical Assembly (Tokyo, October 1975), the 35th World Medical Assembly (Venice, October 1983) and the 41st World Medical Assembly (Hong Kong, 1989) – complemented the Nuremberg Code by introducing the important distinction between therapeutic research (that is research in which investigation is intermingled with the provision of care for the patient/subject in such a way that the research is intended for the direct benefit of the subject as well as for the benefit of society) and nontherapeutic research (that offers no direct benefit to the patient/subject). Like the Nuremberg Code, each version of the Declaration of Helsinki has been accepted by physicians and researchers worldwide as a sound Code of Ethics.

The first of 12 basic principles of research set forth in the Declaration of Helsinki (in the 1975 version) states: 'Biomedical research involving human subjects must conform to generally accepted scientific principles and should be based on adequately performed laboratory and animal experimentation and on a thorough knowledge of the scientific literature.' Like the Nuremberg Code, the Declaration of Helsinki, in stating the first of 12 basic principles, requires the use of animals in research, wherever possible, prior to involvement of human subjects. And also like the Nuremberg Code, the Declaration of Helsinki, including all of its updated versions, makes no exception for any species or subspecies of animal.

If the Nuremberg Code or the 1964 version of the Declaration of Helsinki had intended to limit or exempt some species of animal from the general principle that human experimentation must be preceded, whenever possible, by experimentation involving animals, then surely it would have been clarified in the revisions of 1975, 1983 or 1989. Since the decade of the 1980s has seen the rise of self-styled animal rights activists, it was not possible that those who drafted the last two versions of the document were unaware of the strident opposition to the use of animals, particularly nonhuman primates, as research subjects. Nevertheless the Declaration of Helsinki, though often revised, has remained unchanged in its assertion that research involving human subjects 'should be based on adequately performed laboratory and animal experimentation'. We are therefore safe in concluding that neither the Nuremberg Code nor the Declaration of Helsinki intended to limit or restrict preclinical experimentation to certain species or subspecies of animals.

The Nuremberg Code and the Declaration of Helsinki (as updated) stand as two of the most important and influential documents in medical history. An argument based on an authority is often a weak argument. In this case, however, the two authoritative documents have been scrutinized again and again by scholars in every part of the world. No serious criticism has been leveled against them.

We are safe in claiming that they provide impeccable authority for the statement that it is permissible and often mandatory to use animals, including nonhuman primates, in biomedical research.

The documents not only provide authority for the conclusion that we have drawn, but they provide a powerful rationale for that conclusion.

The very capability of human beings to exercise informed consent by evaluating complex arguments for or against participation in research and making a reasonable choice sets human beings apart from animals, even those animals (nonhuman primates) closest to humans in genetic and physical makeup. This capability to exercise autonomy, along with many other traits that are peculiarly or especially human, establishes in humans an absolute right not to be used by others without informed consent. Animals, even the highest of animals, do not enjoy such a right.

References

1 Permissible Medical Experiments. Trials of War Criminals before the Nuremberg Military Tribunals under Control Council Law No 10: Nuremberg, October 1946 to April 1949. Washington, US Government Printing Office, 1949, vol 2, pp 181–182.

2 Redlich, FC: Medical ethics under national socialism; in Reich WT (ed): Encyclopedia of Bioethics. New York, Free Press, 1978, pp 1015–1020.

3 See discussion of the cases of Salgo versus Stanford University and Natanson versus Kline; in Katz J (ed): Experimentation with Human Beings. New York, Russell Sage Foundation, 1972.

4 Macklin R, Sherwin S: Experimenting on human subjects: Philosophical perspectives. Case Western Reserve Law Rev 1975;25:434–471

5 Bloom SW: The Doctor and His Patient: A Sociological Interpretation. New York, Macmillan Publishing Co, Free Press, 1965.

Charles R. McCarthy, PhD, Office for Protection from Research Risks,
National Institutes of Health, Bldg. 31, Rm 5B59, Bethesda, MD 20892 (USA)

Eder G, Kaiser E, King FA (eds): The Role of the Chimpanzee in Research.
Symp, Vienna 1992. Basel, Karger, 1994, pp 7–17

Changing Cultural and Political Attitudes toward Research with Animals[1]

Larry Horton

Stanford University, Stanford, Calif., USA

At a symposium focusing on the importance of an animal model, it seems particularly appropriate to reflect for a few moments on the politics of animal research.

The entire research enterprise, after all, is heavily dependent on governmental decisions. Government can encourage, actively support, prohibit, discourage, ignore or regulate lines of scientific investigation. It can do so directly and indirectly, and it chooses which course to follow through complex political processes heavily influenced by public opinion and the interplay of various active political participants.

The political health of biomedical research today ought to be judged in fine shape. Government funding for research, though never as much as the scientific community would like, is abundant, and public expectations are high for biomedical science to solve health and environmental problems. But even a healthy system can suffer in parts and has points of stress. The use of animals in research is such a point of stress and political pressure.

The aim of this paper is the following: first, review the highlights of the activism against research with animals in the past decade; second, make observations about the political dynamics of the controversy; third, reflect on the Nuremberg code with respect to animal research; fourth, review some data on public opinion, and finally, conclude with remarks about the future.

[1] The remarks for this symposium present some material not previously published by the author but also draw from two of his previously published articles [1, 2].

The 1980s and Beyond

The 1980s saw the most intense worldwide activism on animal protection issues of any decade in this century. Refreshed in the late 1970s by new writings – most notably Australian philosopher Peter Singer's *Animal Liberation* – a new animal protectionist militancy began to coalesce under the broad banner of animal rights. Animal rights is concerned with all uses of animals by humans – for example for food, entertainment and clothing – but a particularly virulent strain of activism focused on research. This strain manifested itself in many ways:

(1) The 1980s saw the birth of new militant groups, rapid growth in membership of animal protectionist organizations and a decided shift by many formerly moderate animal welfare organizations away from traditional welfare issues and toward opposition to research itself. The movement is well financed, with the combined budgets of animal protectionist organizations estimated to be about $ 50 million [2].

(2) Sensational press coverage of animal rights activism and the transmission of animal rights messages in popular culture – particularly in movies and television – focused public attention on animal issues, particularly on alleged abuses in research. By claiming kinship to other rights movements of this century – civil rights, women's rights and human rights – activists found a niche attractive to the press and popular culture.

(3) Intense political activity erupted in many countries and at many levels. The UK, the FRG, Switzerland, the Netherlands and Italy have passed fundamental new laws governing research with animals. In the USA, restrictive legislation was considered in 25 states and enacted in 10. The Congress passed two important amendments to existing laws governing animal research, and the federal government issued new animal welfare regulations, which, according to the government's own estimate, will cost research institutions over half a billion dollars in capital requirements and several hundred million in annual operating costs [2].

(4) Finally, violence against research laboratories and animal facilities became a visible part of the activism. Facilities were vandalized, equipment damaged, research records confiscated and animals stolen or released.

Political Dynamics

What should one make of all this? Quite frankly, many scientists and physicians find the entire debate boring and without any serious intellectual content. And it is quite true that, if one looks deeply at the arguments in today's debate, one will find that virtually all of the arguments against animal research fall into well-worn patterns and have often been discredited. It is, after all, a very old debate. Opposition to the use of animals developed shortly after the emergence of experimental biological sciences in the 19th century, and today's debate merely continues that which was begun long ago. There has not, in fact, been a fresh political argument on the subject developed during the lifetime of anyone now inhabiting the planet.

Nevertheless, the debate is not to be dismissed. New generations can embrace old ideas with new fervor. The opposition to animal research is utterly dedicated and remains active. During the past decade, opponents of research have clearly gained attention and increased public support. And increased public support means increased leverage in politics. As opinions shift, governmental decisions follow not far behind.

What can we discern from the political dynamics of the new activism that can enable us better to understand and respond to it?

First, the debate is not essentially a scientific one, though the language of the opposition to research often uses the trappings of science; it is fundamentally a moral and political debate.

Second, although the driving force behind animal activism is morally based, the overwhelming political argumentation is pseudoscientific – that is, animal research is alleged to be bad science because of the difficulty of extrapolating from animal models to human models, and animal research is declared unnecessary because nonanimal alternatives are available that will do the job. Both of these notions – in the forms espoused by opponents of research – have been repeatedly disproved, but that has not impaired the staying power of these political arguments.

Third, politics always has a heavy element of opportunism. Single incidents of highly publicized abuses – or perceived abuses – have historically been seized upon and exploited for political purposes. For example, videotapes of the head injury experiments on primates at the University of Pennsylvania have been effectively used by opponents of research to promote political change.

Fourth, no matter how offensive and reprehensible, violence and unlawful acts do not represent a fundamental threat to the continuation of biomedical research. Individual projects may suffer damage, and individuals may be harmed, but it is difficult to imagine society tolerating cessation of research due to what are essentially terrorist acts. There are costs to be sure, but increased expenditures on security for research facilities – like airport security checkpoints – have simply become a fact of contemporary life.

Fifth, outright abolition of animal research is not a serious political threat under reasonably foreseeable conditions, but much serious damage can be done to research by measures short of abolition. Sophisticated animal activists realize that research can be restricted and, in some instances, eliminated de facto by governmental actions that reduce or cut off the supply of animals, increase the burden of regulation to unaffordable or unmanageable levels or adversely affect the conditions necessary for the conduct of research. Most close observers of the politics of biomedical research regard such actions as the greatest threat. And because such restrictions or burdens leave the realm of easily understood politics – animal research versus no animal research – and enter the murky world of the degree of regulation, they cannot always be easily dealt with in political processes.

The Swiss example is illustrative here. In 1985, the Swiss voted on a national ballot initiative that would have flatly banned all animal research and testing in Switzerland. The measure was sponsored and vigorously supported by strong animal rights organizations. Biomedical groups actively opposed it. The measure was soundly defeated, with three quarters of the electorate voting against it.

In February 1992, the Swiss once again went to the polls on an animal research initiative. This time the initiative was sponsored by the old-line, quite respectable Swiss Society for the Protection of Animals, and instead of trying to abolish animal research, it imposed restrictions and subjected research to ambiguous standards. Again, the biomedical community actively opposed the initiative. This measure too was defeated – but this time by the narrow margin of 56 to 44%. Animal activists were encouraged by the result and are continuing their efforts; a new abolitionist initiative was defeated on the Swiss ballot in 1993, and another one embodying restrictions may be on the ballot within the next two years.

Sixth, categories of moderate versus extreme have become distorted. Two New York University sociologists, in a book entitled *The Animal Rights Crusade*, published in 1992, make the following observation: 'There are moderates on both sides of the animal protection debate. But they are usually ignored, as the extremists, driven by fundamental visions, become the most visible actors in the controversy' [3].

This view is precisely the same as that given common currency in the popular press: that the animal rights debate is dominated by extremists on both sides. Yet this fundamental view – presented by both the New York University scholars and the popular press – is profoundly in error.

Biomedical scientists who speak out on the issue do not hold extreme positions; rather, they hold truly moderate positions. It is hard to imagine what an extreme position on the scientific side would be – perhaps, the divine right of the researchers to do whatever they want without any oversight or regulations – but no one holds such positions today, or if they do, they keep them well hidden. The biomedical community accepts and advocates the basic tenets that have previously been urged by moderate spokespersons and which the public supports: preference for the use of nonanimal models whenever possible, minimization of pain and suffering, regulation and inspection by qualified officials and – in the USA – participation by laypersons on animal care committees.

The confusion over who is and who is not moderate in the animal research debate arises largely from two circumstances:

(1) Even if scientists are quite moderate, they nevertheless are invariably at one end of the generally bipolar presentation of the issue. That is, press reports on an animal research issue commonly quote an authentic, admittedly radical animal rights activist on one side of the issue and a distinguished biomedical scientist on the other. Hence, in the normal presentation of the controversy, the Nobel laureate and the antivivisectionist occupy opposite sides, leaving the middle open for self-proclaimed moderates to work compromises.

(2) Many animal protectionists who claim to be moderates actually embrace strong antiresearch positions. Some groups traditionally devoted to animal welfare – including humane societies – have been transformed and have moved beyond animal welfare issues and toward animal rights activism. The public has not caught up with this change.

I emphasize this issue because it may have serious consequences with respect to future political decisions affecting research with animals. If one side of a controversy is actually quite moderate and the other side is verifiably extreme, the process of accommodation in the middle will necessarily result in the sure drift away from moderation. It is of critical importance, therefore, that the true, moderate credentials of the biomedical community be accurately known.

A final observation about the politics of animal research is that both biomedical research and animal protection are seen as distinct social goods worthy of simultaneous and parallel support. One seeks to save lives and alleviate suffering through the advancement of knowledge and the discovery of cures and treatments; the other seeks to care for and protect animals. Both are earnestly valued objectives, both are motivated by compassion and caring, and both appeal to those same qualities in decision makers and in the public.

Much research does not involve animals and can be supported by persons with the strongest commitment to animal rights, and conversely, much in the animal protection movement can be enthusiastically supported by scientists using animals in research. But in view of the current inability of biological science to acquire some essential information by any means other than using animal models, and given that some animal activists believe that any use of animals in research is immoral and must be stopped, a fundamental conflict is inevitable. It is a real conflict, and at some point a stark choice cannot be avoided.

But political conflicts rarely result in stark choices. Politicians do not like dealing with such choices. Neither the public nor politicians want to choose between the two popular and desirable social goods of biomedical research and animal protection. Whenever possible, ways will be sought to support both. It is here that perceptions of reasonableness and moderation will carry great weight. The danger is that what may appear to be a compromise may actually inflict real damage to research.

The Nuremberg Code

With respect to the Nuremberg Code, a slight digression is in order. Spirited rhetoric is expected in radical political movements, but one rhetorical excess is so offensive and so erroneous that it deserves special comment. I refer to the comparison of contemporary laboratory animal research to experiments conducted under Nazi auspices during the Second World War. This has become a staple item in certain parts of the antiresearch movement. Researchers are compared to Nazis and research facilities to death camps. Swastikas are commonly scrawled on protest posters and on vandalized research facilities.

The irony of this situation is twofold: first, Nazi Germany was notoriously hostile to animal research and remains the only modern state ever to have adopted antivivisection as a state policy; second, the Nuremberg Code, a set of ethical principles forged in direct response to Nazi research abuses, specifically embraces animal research.

What we now call the Nuremberg Code is actually part of the judgment of the Court in a 1946–1947 war crimes trial at Nuremberg known as the 'medical case'. Of 23 defendants, 20 were physicians, some quite distinguished. At issue were various experiments subjecting humans to malaria, mustard gas, seawater ingestion, epidemic jaundice, typhus, poisons, burns, high-altitude compression, freezing weather, euthanasia, sterilization and other diseases and traumas. These experiments resulted in appalling suffering and loss of life. Consequently, the prosecution, the defense and the judgment of the court dealt extensively with medical ethics. The attitude of the Nazi state toward animal research was specifically addressed by the prosecution, and testimony and cross-examination dealt with the relationship of animal research to human experimentation.[2]

[2] See the opening statement of the prosecution by Telford Taylor [4] and the testimony of Dr. Andrew Ivy [5, pp. 83–84].

The judgment of the Court, handed down in August 1947, included a discussion of medical ethics and a section on permissible medical experiments. The Court acknowledged that human experiments 'yield results for the good of society that are unprocurable by other methods or means of study. All agree, however, that certain basic principles must be observed in order to satisfy moral, ethical and legal concepts' [5, p. 181].

The Court then enumerated 10 principles that remain to this day the ethical and moral foundations for human experimentation. The first principle is that 'the voluntary consent of the human subject is absolutely essential'. What the Court meant by voluntary consent was spelled out in the accompanying explanatory text that reads like a crisp summary of today's standards of informed consent. The remaining principles set out requirements to avoid all unnecessary physical and mental suffering and injury, to ensure preparations and facilities to protect subjects, to set out the qualifications and responsibilities of those conducting experiments and to establish conditions under which experiments should be discontinued. Importantly, principle number 3 was as follows: 'The experiment should be so designed and based on the results of animal experimentation and a knowledge of the natural history of the disease or other problem under study that anticipated results will justify the performance of the experiment' [5, pp. 181–182]. This determination that research with animals is an ethically necessary precursor to human experimentation has been subsequently validated by the Declaration of Helsinki in 1964, the World Medical Assembly in Tokyo in 1975 and by many subsequent reports of medical and scientific organizations [6].

Public Opinion[3]

A comprehensive public opinion survey conducted in mid-1989 on the attitudes of both adults and children toward research with animals and the opposition to such research provides some interesting data and suggests some conclusions and trends. This survey was conducted in two parts – a survey of 1,000 people 6 years of age and over selected at random from throughout the USA and a survey of 400 randomly selected 9- to 14-year-old children living in the USA.

[3] This section of the symposium remarks was accompanied by 10 slides of data taken from the referenced report. That information is summarized here in prose form. All the data below are from 'Attitudes toward the use of animals in research: A report of findings from nationwide surveys of adults and children', prepared for the American Medical Association, September, 1989 [7].

When given a list of eleven issues, only 9% of the adult respondents thought 'stopping research on animals' was one of the most important issues facing the USA. Only 12% said 'protecting the rights of animals' was such an issue. These two issues ranked at the bottom of the list in terms of their importance. But it would probably be in error to conclude that these issues are not significant ones to the public. The issue that ranked highest in importance was 'improving education', which was selected by only 33%. 'Cleaning up the environment' received 32%, 'reducing the federal deficit' 23% and 'protecting the rights of minorities' 17%. When seen within the context of scattered public opinion on other major issues, the interest shown in animal-related issues is significant – particularly in democratic societies where determined advocates can leverage influence far beyond their numbers.

The adult survey data also show a sharp increase in awareness of animal research issues. In 1982, 35% of a surveyed population answered yes to the following question: 'Have you heard about organizations which are trying to restrict, eliminate or find alternatives to testing on animals?' In 1989, 62% answered yes to that same question.

When asked if they favored or opposed specific uses of animals, adult opinion exhibited a wide range of variance. 96% favored the use of animals as pets, but only 17% favored fur coats (79% opposed fur coats). Other uses fell in between those extremes, with general support for using animals for health-related research and less than majority support for clothing uses or for testing purposes for household products and cosmetics. Interestingly, different wording for the same or similar activity evoked different responses. 79% favor using animals to find cures for cancer and heart disease, 72% for medical research, 71% for testing new medicines, 60% for helping to understand how the body works, 58% for biological research, 56% to find cures for mental illness, 55% for scientific experiments and 49% to find cures for alcohol and drug abuse.

In assessing the credibility of various professions as sources for information and opinion, adults gave the highest rating to scientists and the lowest to political consultants (table 1).

When asked their impressions of various institutions and organizations, the adult survey demonstrated that the public does indeed support both research interests and animal welfare/animal rights interests. The Humane Society received the highest favorable rating: 87% viewed the Humane Society favorably, 6% unfavorably. The next highest rating went to the National Institutes of Health, with 70% favorable and 6% unfavorable. Other ratings were: American Medical Association 79% favorable, 13% unfavorable; animal welfare groups 71/14; People for the Ethical Treatment of Animals 49/12; animal rights groups 64/21, and the Animal Liberation Front 27/17.

Table 1. Adult impressions of various professions (%)

Profession	Favorable	Unfavorable
Scientists	86	5
Doctors	84	12
University professors	74	9
Biomedical researchers	59	9
Accountants	64	12
Lawyers	57	35
Political consultants	34	40

Table 2. Children's attitudes toward using animals in research (%)

	Favor	Oppose	Undecided
Overall	26	53	21
Male	31	45	24
Female	21	61	18
9–10 years old	35	39	25
11–12 years old	21	56	23
13–14 years old	24	58	17

When asked their attitudes toward animals in research and toward animal rights groups, most adult respondents were favorable to both interests – again demonstrating that the public finds both interests worthwhile and supports both activities in parallel. The survey did not pit the interests against one another. Overall, 47% favored research with animals, 33% opposed and 20% were undecided. With respect to the attitudes toward rights groups, 58% favored such groups, 25% opposed and 17% were undecided. When the responses were broken down by age, older persons were found to be more favorably disposed to research with animals than younger persons; conversely, older persons were less favorably disposed to animal rights groups than younger persons. But in all cases, more persons in each age group favored the activity (research with animals and animal rights groups) than opposed it.

Children – aged 9–14 years – were decidedly less favorable than adults toward research with animals and much more favorable toward animal rights groups, with girls showing less favor for research and more for animal rights groups than boys. As shown in this chart, twice as many children opposed using animals in research as those who favor it (table 2).

The children's views on animal rights groups are just reversed. Overall, 54% favor animal rights groups, 24% oppose and 19% were undecided. Sex and age factors here show that girls are more favorably disposed toward animal rights groups than boys – 54% of the boys and 59% of the girls favor animal rights groups while 26 and 22%, respectively, oppose animal rights groups.

Because this survey was conducted at a specific time, it is not known whether the opinions will change as the respondents mature or whether we are seeing a coming shift in public opinion of major proportions. Prudent advocates of all points of view, however, will conclude that public education at all levels should be an important component of any successful advocacy program.

Conclusions

First, some level of intense opposition to the use of animals in research will persist. One can expect multiple political approaches – outright abolition of all research, specific measures to eliminate certain kinds of research or research with certain species, and efforts to restrict or encumber research. And as discussed above, the greatest danger lies in specific measures implemented by the regulatory process; uninformed political decisions could unintentionally damage research.

Second, the key to political success in supporting research rests not with what opponents of research do, but with what the scientific community does. This issue, after all, has been around for a long time, and the current activism is really just the new round of an old fight. The research community's first step must be to ensure that its own house is in order – that is, that facilities are adequate, that high standards of animal care are maintained and that scientific protocols are strictly followed. In addition, the scientific community must engage in public education and political action. Policy makers and the public must understand that research is carefully conducted and that animals are humanely treated. Without public understanding of the necessity to use animals in biomedical research, one cannot expect to retain long-run political support.

Finally, the scientific and health care communities must vigorously support more and better science education. It is well known that opponents of research are taking the long view and are supporting programs aimed at children. We cannot afford to do less.

Despite stresses and potential problems, there is every reason to be optimistic about the outcome of the controversy. If the scientific community understands the potential seriousness of this issue and participates actively, I think the outcome will not be in doubt.

I wish to note in closing that political decisions are always determined by participants in the political process, not spectators. I hope that all of us will be participants.

References

1 Horton L: The enduring animal issue. J Natl Cancer Inst 1989;81:737–743.
2 Horton L: Physicians and the politics of animal research in the 1990s. Cancer Bull 1990;42:211–219.
3 Jasper JM Nelkin D: The Animal Rights Crusdade: The Growth of Moral Protest. New York, Free Press, 1992, p 174.
4 Trials of War Criminals before the Nuremberg Military Tribunals under Control Council Law No 10. Washington: US Government Printing Office, 1949, vol I, p 71.
5 Trials of War Criminals before the Nuremberg Military Tribunals under Control Council Law No 10. Washington: US Government Printing Office, 1949, vol II, pp 181–182.
6 Silverman WA: Human Experimentation: A Guided Step into the Unknown. Oxford, Oxford University Press, 1985, p 155.
7 Unpublished report: Attitudes toward the use of animals in research: A report of findings from nationwide surveys of adults and children. Prepared for the American Medical Association by Mellman and Lazarus, Inc, September 1989.

Larry Horton, MA, Stanford University, Building 170, Stanford, CA 93405–2040 (USA)

Eder G, Kaiser E, King FA (eds): The Role of the Chimpanzee in Research.
Symp, Vienna 1992. Basel, Karger, 1994, pp 18–25

Ethical Aspects of Animal Research

Carl Cohen

The University of Michigan, Ann Arbor, Mich., USA

The scientific work that will be done at the Primate Center of Immuno AG will serve human health and well-being. Those who work here will have other motivations as well, no doubt – personal and economic – but the larger objectives that justify this work are medical and moral. With those moral objectives in mind, I would reflect with you, for a short while, upon the ethical aspects of research using nonhuman animal subjects.

Three items are on my agenda:

First, I will recapitulate very briefly the principal arguments commonly presented as moral objections to animal research and respond concisely to those objections.

Second, I will consider a third objection to animal research, one that is less commonly heard but is of special concern to those who work with primates and care for them – and I will respond to this objection also.

Third, I will argue that research using animals is not merely morally permissible, but that it is entirely right and for some persons and some institutions a strong moral obligation.

Principal Objections

The commonly expressed objections to the use of animals in biomedical research are in essence of two kinds [1]:

(A) First comes that family of objections based upon the conviction that research using animals *violates the rights of animals*. Conduct that violates the rights of others, if it is voluntary and deliberate, is morally wrong. Therefore (the critic contends) all research using animal subjects is ethically wrong and ought to be forbidden, stopped.

Arguments of this kind are usually not intended merely to insure the humane treatment of animal subjects; they are objections to *all* uses of animals, including all uses of them in biomedical research. Those who present such criticism, as they themselves say, do not seek larger cages, but empty cages [2].

Argumentation of this kind is profoundly mistaken. It relies upon a confused conception of rights. It applies to the world of animals moral concepts, and an ethical framework that is in truth intelligible, coherent, only in a human, moral community. The extended response to this argument I have presented in detail elsewhere; very briefly now I indicate the thrust of that response.

We have obligations to animals, of course. But it is a mistake to infer, from the fact that we have an obligation, that the subject of that obligation has a *right* against us. Many of the obligations that we all have – to our students, friends, family, pets – are not grounded in the rights of those students, friends and so on but arise from our commitments to them, and from the special relations between us and them.

For every right there is, indeed, a correlative obligation. If you have a right against me, or us, then I have (or we) have the obligation to respect that right. But the moral proposition that all rights entail the obligations of others cannot be converted simply. Indeed, the converse is false. Rights and obligations are not reciprocals; the relations between them are not symmetrical. And confusion on this point is unhappily very common. It is the mistaken supposition that wherever there is a genuine obligation there must be some right held by another that leads zealous critics to hold that rats have rights, and fishes, and perhaps trees, and whatever else may deserve our protection in the interest of other humans.

Rights – claims or potential claims of one moral agent against another – are intelligible only within a community of beings who recognize general principles of duty as distinct from interest, who can apply general principles of duty autonomously, and who are therefore capable of recognizing the internal aspects of a voluntary act – its intent, and the moral qualities that can arise only from that intent.

Therefore, with full understanding of the sentience of animals and a full appreciation of our obligations to care assiduously for them, I would insist, quite plainly, that *animals do not have rights.* They have interests, of course; but not every interest yields a right. Animals have feelings and suffer pain, of course. But sentience in itself yields no rights. The obligations we have to animals arise partly because they can feel pain and distress and we have a duty, as moral agents, not to cause needless pain. But it is an error – a serious and dangerous error – to conclude from the physical interests of rodents, or primates, that they are moral agents as humans are moral agents.

It is a profound moral wrong for one human being to kill another, in most circumstances. Humans have natural rights, as humans, which we would be morally wrong to infringe. This is not a matter only of the law of the state; outrageous conduct even in the name of the state may be, as we correctly say, 'crimes against humanity'. And we will prosecute such crimes. But animals kill and eat one another constantly; theirs is a world of predation, and in that world there is no moral fault in the predator.

In sum: arguments against the use of animals in research, based upon the violation of their alleged rights, arise from a mistaken application of human moral concepts to a realm in which they do not apply.

(B) The second set of objections to the use of animals in biomedical research abandons reliance upon alleged animal rights. Putting aside all talk about rights, these arguments are founded upon *utilitarian* principles. The pain and distress that we inflict upon animals (say they) are excessive and unwarranted. If utility be the standard of moral judgment (these critics conclude), research using animals cannot be justified [3].

Arguments of this form, in all their varieties, also fail utterly, for two reasons:

(1) First, such an argument mistakenly assumes that the pains inflicted on an animal in research are to be weighed, in the moral calculus, as equivalent to the pains of a human. The argument fails to distinguish the differing moral status of the beings involved. Between the killing of a human child and the killing of a mouse there is the very greatest moral difference, you and I will agree. But this critic finds those who make this distinction to be *speciesists* – a name that (because in English it plays upon the terms 'racist' and 'sexist') sounds highly disreputable. But we here are all speciesists, of course. I put the matter bluntly: all those (in or out of the research community) who are not speciesists, who fail to make needed moral distinctions among species will also almost certainly fail to fulfill their real duties to the members of the human species.

(2) Second, utilitarian objections are profoundly mistaken because they do not correctly weigh *all* of the consequences, good and bad, of using animals in research and not using them. Killing or hurting animals is surely a moral negative. But sometimes that negative is an unavoidable by-product of work that has very wonderful consequences. The general appraisal of research using animals, if it is to be fair and rational, must weigh thoughtfully all that has been accomplished, and all that can only be accomplished, through their use.

Some utilitarian critics are simply ignorant of the central role of animal research in medicine. That ignorance, unhappily very widespread, is itself morally blameworthy when it underlies irrational attacks upon biomedical research. But the moral blameworthiness of critics who are not ignorant is yet greater. There are those who, sincere in their passion to protect animals against pain, refuse to acknowledge publicly what they know privately about the reasons for the uses of animals, who repeat sensational claims about animal research that they know to be false and who sometimes make accusations against animal research that they know, or ought to know, to be unjustifiable.

In sum, on any thoughtful utilitarian analysis, the moral argument against animal research collapses completely. Indeed, the conclusion to which careful utilitarian reasoning compels us is that research using animals is a great good for humankind and is morally right.

I have briefly addressed the argument from rights, and the argument from utility. That completes the first item on my agenda.

Objection to Research with Primates

A third argument against animal research is of particular concern to those whose research involves the use of monkeys and apes. How might one attack the conclusion that primate research is justifiable because its benefits so greatly outweigh its evils?

The attack begins with the premise that no sharp line can be drawn between humans and the lower animals. There is a *continuity* among all the species of animal life; that is a conclusion of evolutionary science that cannot be denied. Because of this continuity of species (the argument continues) we must weigh the evils done to animals in research *on the same scale of values* that we use in weighing evils done to humans. Therefore even rats and rabbits cannot be incommensurably inferior to humans. And between humans and primates from which homo sapiens evolved, *moral commensurability* (this critic continues) is inescapable.

On this foundation the following argument is constructed:

If all species of animals lie on a great continuum, then, if it were morally permissible to exploit any group of animals on the basis of the benefits that their use produces, it must be equally permissible to exploit humans, who are morally commensurable, in the same way.

But it is plainly not morally permissible to exploit humans as we exploit animal lives.

Therefore (the argument concludes) it must be the case that animal research cannot be consistently justified on the basis of the benefits it produces.

The argument is valid in form: If P then Q. Not Q. Therefore not P. (modus tollens is its name.) But I contend the conclusion of this argument is false. And if the argument is formally valid, and yet its conclusion is false, one of its premises must also be false. Which one? That human subjects may not be exploited as animal subjects are is surely true. But the major premise of the argument is not true. It is not the case that, because humans and other animals are commensurable on some scales, we may never do to one group what we may not do to the other. This is an appealing but misleading blunder, a mistake on two levels.

First, suppose we were to grant (arguendo) that there is some single scale for evaluating the evils done to humans and to rats, and to all species on the phylogenetic continuum. It simply does not follow that species falling at different points on that continuum must all be treated similarly or given equal consideration. Differences that are quantifiable really matter. Consider: rich people and poor people differ on one commensurable scale, that of money, but in monetary matters there is no moral obligation to treat rich and poor alike. On the contrary, we think it right to tax the rich as we would not tax the poor, and we would think it foolish to suppose that monetary commensurability forbids us to differentiate in this way. Why? Because important quantitative differences often require different moral judgments, even when those differences are along a single scale. The argument here seeks to infer too much from the alleged 'commensurability'.

And second, the single scale supposed (arguendo) just above, should not be granted in fact. This supposed commensurability is a fiction. There are great organic similarities between humans and other species, of course. When the relations of humans and apes are at issue, great organic similarities will certainly not be denied. These relations we investigate with untiring interest, and they make of primate research a tool of special and extraordinary value, a value so great that we gather here in Vienna today to celebrate the coming intensification of those investigations. Nor is it any wonder that the chimpanzee, of all creatures, should be so attractive and so useful to us as a research subject. But organic continuities do not establish a moral commensurability between humans and chimpanzees.

The central ethical question is this: are the similarities of humans with lower species of such a kind that what is plainly immoral to do to humans must also be immoral to do to animals? The answer is no. They are not because overriding those similarities are profound differences among species, differences that make it an ethical mistake to treat them on the same moral scale.

The critic then rejoins: 'But the differences are nevertheless natural; they have a foundation in our natural history.' And that is no doubt true – but differences they remain nevertheless. To say of humans that they have a moral status fundamentally *distinct* from other primates is not to say that this distinctness must rest upon some miraculous or supernatural endowment. Divine intervention is possible; we need not dispute that. But supernatural causes may be put entirely aside. Whatever the explanation of the development of moral differences among the several species may be, those moral differences are a reality, not to be denied. Those differences make a difference in our judgments and in our conduct.

Three illustrations of the consequences of these differences are these:

(1) We may not experiment upon humans without their freely given and fully informed *consent*. But the moral status of animals is such that, although their organs be similar, and although they be more or less compliant, the quest for their informed and voluntary consent to cooperate in the achievement of the ends of the investigation is an absurdity.

(2) Humans *rightly* do to animals what animals cannot have the *right* to do to us. We exterminate roaches and rats. Now sometimes rats eat human babies, but they have no right to do so; indeed they are totally incapable even of conceiving of any moral status for any of the things that they do. Rights (as noted earlier) are not the kinds of things that rats, or other animals, can be intelligibly said to possess.

(3) Humans can do *wrongs* that animals cannot do. An act is culpable, blameworthy, when the deed is done with a guilty mind. In very primitive societies animals were sometimes punished for crimes – but we understand full well that no rat or monkey ever committed a crime, because animals by nature do not possess the state of *moral awareness* necessary for the attribution of iniquity or wrongdoing.

The argument from commensurability reduces, in the end, to a version of the argument based upon animal rights. But the moral divide between humans and animals (whatever its natural history) is vast and unbridgeable. The argument against animal research based upon alleged 'scientific' continuities among species is, therefore, without merit.

This completes the second element on my agenda.

Moral Obligation to Research Using Animals

I submit that research using animal subjects is not merely ethically permissible and right but is in some circumstances a *duty*. It is a duty for those who commit themselves, professionally and publicly, to the care and well-being of humans. And it is a duty that falls with special weight upon commercial firms that profit from the production of drugs used in the care of humans.

The benefits that accrue to all of us from the sale and use of pharmaceuticals carry with them some weighty obligations to the community. Two of these obligations I register here:

(1) We are morally obliged to determine, with the highest reasonable degree of scientific reliability, that the drugs we put forward for clinical trials with

humans are *safe, are not toxic.* Toxicological studies are therefore a major activity and a principal duty of every pharmaceutical company. Without the use of the animals that are absolutely central in toxicological studies, we would fail, and fail irresponsibly, in our duty to protect human subjects in subsequent investigations so far as we are able.

(2) We are morally obliged to determine, with the highest reasonable degree of scientific reliability, that the drugs we put forward for clinical trials with humans are *efficacious* – that they do what we expect them to do, or at least that our evidence to support that judgment is the best that it is in our power to get. To this end the use of animals is not simply a possible way to proceed. It is, in many if not most cases, the only way to acquire reliable evidence of efficacy before human subjects become involved. It is true of course that, animal studies concluded, we cannot be completely confident of the effectiveness of any drug until trials with humans have been run. But if we did not engage in animal studies as preparation for those trials we would certainly fail to fulfill our moral obligations to humans.

Without some good evidence of efficacy we could not justifiably put human research subjects at risk. Refraining from animal studies in the investigation of efficacy, therefore, must result in our inability to test, to market and then to prescribe drugs that may eventually prove of enormous value to humankind.

It is this point, I believe, that so many in the general population of North America and Europe simply have not grasped. There are millions who genuinely care for animals, who readily grant our higher duties to humans – but who simply do not understand the central role of animal subjects in pharmaceutical research. I recapitulate here what every person at this gathering knows well, and what all who think about research with animals ought to weigh with care:

In testing the efficacy of a new drug, *there is no replacement for trials using live organisms.* Tissue samples, chemical analyses, computer simulations – none of these, or anything else, can replace the living organic being in research whose accurate results directly affect human lives. Therefore, when first we administer any new drug to a live organism, that administration *must* be an experiment. We will not distribute (or be permitted to distribute) new drugs to the general public without the most scrupulous experimentation in advance. The risks of that experimentation cannot be imposed upon human subjects without preceding animal studies. Therefore, as a society, as a culture, we face this absolutely inescapable dilemma: we will do toxicological studies and efficacy studies on animals in developing new drugs – or we will not develop those new drugs. Anyone who believes that it is not merely our interest but our duty to develop new drugs to advance human well-being must also agree that research on those drugs using animals is a moral necessity. Therefore, the failure to use animal subjects in biomedical research must result in a failure, by some, to fulfill their moral duties.

References

1 Cohen C: The case for the use of animals in biomedical research. N Engl J Med 1986;315:865–870.
2 Regan T: The Case for Animal Rights. Berkeley, University of California Press, 1983.
3 Singer P: Animal Liberation. New York, Avon Books, 1975.

Carl Cohen, PhD, Residential College and Medical School,
Program in Human Values in Medicine, The University of Michigan,
3269 Medical Science I, Ann Arbor, MI 48109–0624 (USA)

Eder G, Kaiser E, King FA (eds): The Role of the Chimpanzee in Research.
Symp, Vienna 1992. Basel, Karger, 1994, pp 26–33

Animal Welfare Regulations and Accreditation by the American Association for Accreditation of Laboratory Animal Care

Impact on Chimpanzees in Research

Harry Rozmiarek

Office of University Laboratory Animal Resources, University of Pennsylvania, Philadelphia, Pa., USA

Animals have been used in the laboratory and classroom for hundreds of years, and most early knowledge of anatomy, physiology and medicine came from observations and studies of animals and animal systems. These discoveries were basic and primitive compared with current research, but they formed the foundation of knowledge upon which further discoveries and scientific and medical advances continue to be based. The care and concern for the well-being of laboratory animals has progressed during this time from early concerns for animal life and maintenance to current understanding of adequate nutrition, latent disease hazards, genetic makeup and psychological behavior patterns. Animal holding facilities have evolved from back rooms and garages to sophisticated barrier facilities with filtered air and climate-controlled environments, while the concern for animal well-being now includes a real effort to provide psychological well-being as well as physical requirements.

This discussion will outline some of the major animal welfare policies and regulations in the USA [1] and how they are interpreted and implemented. These policies range from provisions required by law to voluntary accreditation programs in which many institutions participate. Individual institutions and facilities also have local and regional policies that supplement broader provisions and are tailored to their specific operations. Programs of environmental enrichment enhance the well-being of laboratory animals but are often complicated by requirements for separation of incompatible animals and containment of zoonotic agents and biohazards of natural or experimental nature.

Legislation

One of the earliest pieces of legislation that directly addressed laboratory animals was enacted in the UK in 1822 and protected animals used in research. The English Cruelty to Animals Act was established in 1876 and has remained in effect. This act permits animal experimentation only under license from the Home Secretary. It also requires various certificates in addition to a license for scientists to carry out certain procedures.

The first law in the USA protecting nonfarm animals, commonly known as the Laboratory Animal Welfare Act, was established as Public Law 89–544 in 1966. It regulated trade in dogs and cats procured for laboratory research, as well as dogs, cats, hamsters, guinea pigs, rabbits and nonhuman primates held by certain research facilities. Amendments to the act over the years have titled it the 'Animal Welfare Act' [2], and it now specifically addresses exercise for dogs, a physical environment to promote the psychological well-being of nonhuman primates and details about the organization and functioning of institutional animal care and use committees in addition to its more basic provisions covering animal care. A recent change in policy by the US Department of Agriculture (USDA) now enforces the Act using 'performance standards' by veterinary field inspectors in their regular inspection of facilities in addition to registration of research facilities and licensing of laboratory animal dealers. While the Act includes all warm-blooded animals, the implementing regulations [3] exclude birds, domestic rats and mice, horses and other farm animals used for food, fiber or nutrition and production research. These interpretations by the USDA have been challenged in US federal court by animal rights organizations and are currently under discussion. The Act also requires that 'each research facility shall provide for the training of scientists, animal technicians and other personnel involved with animal care and treatment' and must include 'instruction on research or testing methods that minimize or eliminate the use of animals or limit animal pain or distress'. Each research facility must appoint an Institutional Animal Care Committee (IACUC) to assess the research facility's animal program, facilities and procedures. This committee, composed of a chairman and at least two additional members, includes a Doctor of Veterinary Medicine with training or experience in laboratory animal science and medicine, and a member who is not affiliated in any way with the facility to represent the general community interests in the proper care and treatment of animals. Semiannual written reports of these evaluations are presented to the responsible institutional official and are made available to officials of funding federal agencies for inspection and copying upon request. Significant deficiencies which cannot be corrected in adherence with the plan and schedule are reported to the USDA as well as to any federal agency providing funds.

Guidelines

While the Animal Welfare Act provides that minimal laboratory animal care requirements must be met, further policies and guidelines provide direction and recommendations to assist institutions in caring for and using laboratory animals in ways judged to be professionally and humanely appropriate. The best known and most widely used of these in the USA is the *Guide for the Care and Use of Laboratory Animals* [4]. This document was first published in 1963 under the title *Guide for Laboratory Animal Facilities and Care* and was revised in 1965, 1968, 1972, 1978 and 1985. The Guide covers virtually every aspect of the care and use of laboratory animals, beginning with institutional policies for monitoring the care and use of animals and assuring that professionals caring for and using animals are properly qualified. Each institution is encouraged to establish an animal care and use program that is appropriately staffed and managed, and a committee to monitor the program. A veterinarian should always be associated with the program, preferably one with training and experience in laboratory animal medicine. Close physical restraint of animals is discouraged, and multiple major survival surgical procedures on a single animal are prohibited unless specifically reviewed and justified. Recovery surgery in all animals must be conducted aseptically, and devoted sterile surgical suites must be provided for all nonrodent animal surgery.

Husbandry requires an appropriate social as well as physical environment. Space recommendations that equal or exceed those of the USDA are made for each species, with strong encouragement for use of pens and runs rather than cages. Provision for exercise is required, especially for animals to be held for long periods. Attention is given to the microenvironment and the thermoneutral zones of various species in providing appropriate temperatures, humidity and air quality. Lighting and noise levels are considered, and specific recommendations are made for various species. Food, water and sanitation practices must be appropriate for the species. All animals should be identified, and individual health records are encouraged for dogs, cats, primates and farm animals. Provisions for treatment of sick animals, quarantine, isolation and separation of species are recommended to reduce both disease transmission and interspecies conflict. Euthanasia must be performed rapidly and painlessly and should comply with recommendations of the American Veterinary Medical Association Panel on Euthanasia [5]. Specific needs of farm animals used in biomedical research are included, and guidelines for housing, food and care are provided.

A two-volume publication entitled *Guide to the Care and Use of Experimental Animals* [6] has been prepared and published by the Canadian Council on Animal Care. This covers some of the subject material in more detail than the American Guide and is recommended and used by the Canadian Council in their assessment site visits to animal care and use facilities. Similar guidelines have been published by a number of biomedical institutions and associations and provide additional information for animal care and use in different research situations [7–15].

Public Health Service Policy

If an institution wishes to receive federal research funds, it must comply with all Public Health Service (PHS) animal care policies and principles. While the primary PHS reference has been the *Guide for the Care and Use of Laboratory Animals* (Guide), supported institutions must also comply with the Animal Welfare Act and a number of other federal, state [16] and local laws and statutes. In 1985, the PHS reviewed and updated their policy. The new policy, the *Public Health Service Policy on Humane Care and Use of Laboratory Animals by Awardee Institutions* [17], is more comprehensive and complete than before and has made an impact upon the biomedical community. In addition to endorsing the Guide, the 'US Government Principles for the Utilization and Care of Vertebrate Animals Used in Testing, Research and Training' and the Animal Welfare Act, the policy provides for the implementation and supplementation of these principles. This is done by written assurances to the PHS, which are reviewed and updated annually, the appointment of an IACUC that actively oversees the institutions' animal programs, facilities and procedures and the review of animal care and use portions of each application and proposal submitted to the PHS.

For a proposal involving the use of animals to be considered for funding by the PHS, it must originate from an institution that provides an acceptable level of animal welfare assurance. This assurance is submitted to the PHS in writing and may be accepted for a period of up to 5 years. It must include a complete description of the animal care and use program, including descriptions of the physical facilities, administrative authority and responsibility lines, provisions for veterinary care and the IACUC makeup and functions. The institution must categorize itself as either being accredited by the American Association for Accreditation of Laboratory Animal Care (AAALAC) or must evaluate its programs and facilities by some other means.

If the institution is not AAALAC accredited, the IACUC must evaluate all facets of the program and prepare a written report at least once a year. This report must be made available to the Office for the Protection of Research Risks (OPRR) at any time upon request and must describe the nature and extent of the institution's adherence to the Guide and PHS policy. Any deficiencies must be classified as major or minor, must be described, and a reasonable and specific plan and schedule for correcting each deficiency must be included. The OPRR will receive and evaluate the initial report and make decisions as to its evaluation and acceptance. It will also monitor and evaluate continuing acceptability of institutional assurances, which may include unannounced site visits to the institution to aid in this evaluation.

Animal Care and Use Committee

To comply with PHS policy, each institution must appoint IACUC [18] of no fewer than five members with responsibilities similar to the committee required by the Animal Welfare Act. The committee must include at least one Doctor of Veterinary Medicine with training or experience in animal research, one practicing scientist with experience in animal research, one member whose primary interests are in a nonscientific area and one individual who is not otherwise affiliated with the institution in any way. The committee may, and probably will, include more than the mandatory five members and may include other categories of expertise as well. The committee is responsible for reviewing and providing the animal welfare assurances for the institution and, in that capacity, reports directly to the PHS. Their activities include a review of humane care and use programs at the institution at least twice a year and an inspection of all institution and satellite animal holding facilities. They review concerns involving the care and use of animals at the institution, make recommendations to the institution's responsible official and have the authority to suspend any activity involving animals that does not meet the PHS policy. If the committee takes such action, the institution shall review the reasons for suspension, take appropriate corrective action and report that action to the OPRR.

The IACUC is responsible for the review of all proposals or applications using animals which are submitted to the PHS for consideration for funding. It reviews all sections related to the care and use of animals to determine if they are in compliance with PHS policy, including the Animal Welfare Act and

Guide provisions. Outside consultants may be used by the committee to provide expert opinion as required. The committee has authority to approve or withhold approval for each proposal and will notify the investigator and institution in writing of its decision. If a proposal is disapproved, specific reasons will be stated. The PHS must be in receipt of verification of approval for each proposal involving animals by the IACUC before that proposal will be considered.

Whereas many biomedical research, teaching and testing institutions have had effective mechanisms to assure appropriate animal care and use in place for many years, the new PHS policy requires that all supported institutions will use similar mechanisms and follow similar levels of compliance. It also requires that deficient institutions will be brought into compliance quickly and does so with a minimum of oversight from outside the institution.

The Food and Drug Administration, NIH and the USDA's Animal and Plant Health Inspection Service have jointly adopted an interagency agreement concerning their respective responsibilities for regulating animal care and use. Each agency has agreed to share lists of establishments that fall under the purview of each agency, exchange lists of establishments at least quarterly that have been inspected or visited by each agency, exchange information on 'significant adverse findings' revealed by such inspections and inform each other of evidence of serious noncompliance with standards for the care and use of laboratory animals. A majority of institutions that use laboratory animals fall under the purview of one of these agencies, and compliance with acceptable standards of animal care and use is assured.

AAALAC Accreditation

In 1965 a group of biomedical institutions supported the founding of a group called the AAALAC (Bethesda, Md., USA). Using the Guide and Animal Welfare Act as its primary reference documents, the AAALAC functions as an independent accrediting agency. A Board of Trustees representing each of the 36 member organizations serves as a governing and policy-making body for the Association. The working body is the Council on Accreditation and is made up of 20 experts who visit institutions on a regular 3-year cycle to evaluate animal care and use programs for compliance with AAALAC accreditation standards. The purpose of the AAALAC is to provide a program for the accreditation of laboratory animal care and use that will assure the welfare of laboratory animals and enhance the quality of scientific research using experimental animals. Participation in this peer evaluation program is voluntary, and, to date, more than 3,000 site visits have been conducted on three continents, and fully accredited programs have been located in five countries. Fully accredited programs include: universities; schools of medicine, veterinary medicine, dentistry, pharmacy, biological science and nursing; hospitals; pharmaceutical manufacturers; Veterans Administration Medical centers; government laboratories; commercial laboratories; nonprofit research laboratories, and laboratory animal breeders. On-site evaluations are made regularly to evaluate all segments of the animal care and use program. Site visitors evaluate physical facilities as well as programmatic aspects of animal use and care, and visit research laboratories and surgical suites to discuss the use of animals with principal investigators. Accreditation by the AAALAC is testimony that an institution meets or exceeds all current standards of good laboratory animal care and use.

Psychological Well-Being of Nonhuman Primates

While regulations and guidelines offer specific recommendations for the minimum cage space required, the provision of an adequate total environment for laboratory animals is a much more difficult issue and requires considerable professional interpretation. Providing a 'physical environment to promote the psychological well-being of nonhuman primates' when animals are confined to laboratory facilities is exceedingly complex and requires a great deal of ingenuity and constant attention. This is especially true for the curious, intelligent and sometimes strongwilled chimpanzee. Work with hazardous biological agents and the proximity to urban populations further complicate this use. Physical space for a chimpanzee has been defined as over 25 ft^2 per animal with sufficient height and equipment for normal postural adjustments which include brachiation, and this is regularly provided and exceeded. Efforts to enrich and embellish this area include toys, food and other rewards for accomplishing manipulative tasks, opportunities for foraging and direct interaction through pair and group housing. Visual and sensory contact is often provided when compatibility or safety prohibit direct contact. The *NIH Nonhuman Primate Management Plan* [19] includes many published accounts of enrichment programs, and most are still under evaluation.

A good animal care and use program must begin with a clear understanding of the requirement and needs of animals in the program, and a sincere and consistent effort to provide for those needs as completely as possible. We must continue to explore new methods of housing and psychological enrichment as well as less invasive and even alternative methods to the use of living animals as we balance the fiscal and ethical costs of using animals in research with the benefits provided in the advancement of science and medicine. The quality of the entire program and the benefits of the work being conducted should be regularly and clearly communicated to the public.

References

1 Rozmiarek H: Current and future policies regarding laboratory animal welfare. Invest Radiol 1986;22:175–179.
2 Animal Welfare Act and Amendments. Public Laws 89–544, 91–579, 94–279, 99–198, USA.
3 Provisions of the Animal Welfare Act. Code of Federal Regulations. Title 9, subchapter A, parts 1, 2 and 3, USA.
4 Guide for the Care and Use Laboratory Animals. NIH Publ No 85–23. Washington, US Department of Health and Human Services, 1985.
5 1993 Report of the AVMA Panel on Euthanasia. J Am Vet Med Assoc 1993;202:229–249.
6 Guide to the Care and Use of Experimental Animals. Ottawa, Canadian Council on Animal Care, 1980, vol I and II.
7 Guide for the Care and Use of Agriculture Animals in Agricultural Research and Teaching. Washington, Division of Agriculture, NASULGC, 1988.
8 Guidelines for Ethical Conduct in the Care and Use of Animals. Washington, American Psychological Association, 1991.
9 Institutional Administrators' Manual for Laboratory Animal Care and Use. NIH Publ No 88-2959. Washington, US Department of Health and Human Services, 1988.
10 Use of Laboratory Animals in Biomedical and Behavioral Research. Washington, National Research Council, National Academy Press, 1988.
11 Principles and Guidelines for the Use of Animals in Precollege Education. Washington, Institution for Laboratory Animal Resources, 1989.
12 Recommendations for Governance and Management of Institutional Animal Resources. Washington, Association of American Medical Colleges and the Association of American Universities, 1985.
13 Preparation and Maintenance of Higher Mammals during Neuroscience Experiments. NIH Publ No 91–3207. Washington, US Department of Health and Human Services, 1991.
14 Guidelines on the Care of Laboratory Animals and Their Use for Scientific Purposes. South Mimms, Potters Bar, Universities Federation for Animal Welfare, 1989.
15 Live Animal Regulations. Montreal, International Air Transport Association, 1986.
16 State Laws Concerning the Use of Animals in Research. Washington, National Association for Biomedical Research, 1987.
17 Public Health Service Policy on Humane Care and Use of Laboratory Animals by Awardee Institutions. Bethesda, Office for Protection from Research Risks, 1985.
18 Institutional Animal Care and Use Committee Guidebook. NIH Publ No 92–3415. Washington, US Department of Health and Human Services, 1992.
19 National Institutes of Health Nonhuman Primate Management Plan. Bethesda, Office of Animal Care and Use, 1992.

Harry Rozmiarek, DVM, PhD, Office of University Laboratory Animal Resources, University of Pennsylvania, 100 Blockley Hall, Philadelphia, PA 19104-6021 (USA)

Eder G, Kaiser E, King FA (eds): The Role of the Chimpanzee in Research.
Symp, Vienna 1992. Basel, Karger, 1994, pp 34–35

The Human Genome Project

Leroy Hood

NSF Science & Technology, Center for Molecular Biotechnology,
California Institute of Technology, Pasadena, Calif., USA

The Human Genome Project [1] has two major goals: (1) to develop more powerful tools for handling, mapping, sequencing and analyzing DNA and (2) to apply these tools to obtain detailed physical and genetic maps of the genome and ultimately the entire DNA sequence. My view is that we need tools at least 100-fold more effective than those now available before we should mount a large-scale attack on sequencing the entire genome. Perhaps it will take 10 years to achieve this objective of technology development, and during that time the magnitude of mapping and sequencing efforts should be limited by the power of the tools at hand. Moreover, technology development and sequencing efforts should go hand in hand. The areas chosen for large-scale DNA sequence analysis [2, 3] should have as one objective driving technology development. The loci chosen for analysis should be of significant biological and genetic interest so that as sequence information is obtained, it can be applied directly to exciting biology or genetics. Finally, homologous regions of the human and mouse (or primate) genomes should be sequenced to identify the conserved regions that reflect the consequences of natural selection in preserving information representing genes, regulatory regions or features of chromosomes responsible for their special functions (e.g. replication, compaction, segregation) [4–7]. We have chosen to map and eventually sequence the three loci for T cell receptors (α/δ, β, γ) in both the mouse and human. These loci range in size from 150 kb to 2 mb. These loci are rich in genetic and biological information. Currently our mapping efforts are focused on the murine α/δ locus (–1 mb) and the human β locus (–1 mb). Our sequencing efforts are focused on the human β locus and the human and mouse α/δ loci. I will review our efforts to develop new mapping, sequencing and computational strategies for this effort as well as pointing out the challenges that remain for the future. I will also illustrate the power of the information that is derived from these large-scale sequencing efforts.

References

1 Hood L, Hunkapiller T, Solomon J: Computational problems and the human genome project. Proc 5th SIAM Conf on Parallel Processing for Scientific Computing, 1991.

2 Hunkapiller T, Kaiser RJ, Koop BF, Hood L: Large-scale DNA sequencing. Curr Opin Biotechnol 1991;2:92–101.

3 Hunkapiller T, Kaiser RJ, Koop BF, Hood L: Large-scale and automated DNA sequence analysis. Science 1991;254:59–67.

4 Koop BF, Wilson RK, Wang K, Vernooij B, Zaller D, Kuo CL, Seto D, Toda M, Hood L: Oranization, structure and function of 95 kp of DNA spanning the murine T-cell receptor Cα/Cγ region. Genomics 1992;13:1209–1230.

5 Nickerson DA, Kaiser R, Lappin S, Stewart J, Hood L, Landegren U: Automated DNA diagnostics using an ELISA-based oligonucleotide ligation assay. Proc Natl Acad Sci USA 1990;87:8923–8927.

6 Wilson RK, Hood L: High-throughput fluorescent DNA sequence analysis: Methods and automation. Methods (A Companion to Methods Enzymol) 1991;3:48–54.

7 Wilson RK, Koop BF, Chen C, Halloran N, Sciammis R, Hood L: Nucleotide sequence analysis of the 3′ terminal region of the murine T-cell receptor α/δ chain locus: Strategy and methodology. Genomics 1992;13:1198–1208.

Leroy Hood, PhD, NSF Science & Technology, Center for Molecular Biotechnology,
California Institute of Technology, Division of Biology, 139–174, Pasadena, CA 91125 (USA)

Eder G, Kaiser E, King FA (eds): The Role of the Chimpanzee in Research.
Symp, Vienna 1992. Basel, Karger, 1994, pp 36–42

Man and Chimpanzee: An Evaluation of Genetic Similarities

Héctor N. Seuánez

Genetics Section, Instituto Nacional do Cancer, and Department of Genetics, Universidade Federal do Rio de Janeiro, Brazil

Genetic studies have been most valuable in showing remarkable similarities between man and the chimpanzee (*Pan troglodytes*) across a wide spectrum of biological complexity. While the average chimpanzee protein has been found to share 99% of amino acid homology with that of man [1], molecular comparisons have consistently confirmed that humans and chimpanzees share a substantial proportion of almost identical genes, both at their nuclear and mitochondrial genomes. Within the nuclear genome, these include single-copy sequences [2, 3] and different repetitive components, such as satellite fractions [4, 5], ribosomal gene multifamilies [6] and sequences of retroviral origin [7]. Furthermore, karyological comparisons between man and the chimpanzee show that these species share relict chromosome homologies (homoeologies) due to their evolutionary conservation since the time they split from an extinguished common ancestor [8, 9]. These similarities, when considered within the logical framework of the evolutionary theory, are good evidence that biological differences between humans and chimpanzees are continuous rather than discrete. Furthermore, comparative analyses of human-chimpanzee psycholinguistic characteristics have confirmed that differences between the two species are a matter of degree rather than a matter of kind. Thus, the striking anatomical traits that so clearly distinguish us from the chimpanzee remain unexplained despite our efforts to interpret them at a level of molecular complexity.

Biochemical studies, in fact, point out that humans and chimpanzees resemble one another as sibling species of other organisms [1]. It might be argued that electrophoretic analyses of protein products explore, indirectly, only a small component of the genome, mainly a fraction of the DNA that is involved in the process of transcription/translation. Moreover, as this methodology might overlook samesense mutations resulting from the degeneracy of the genetic code or missense mutations resulting in amino acid substitutions of similar charge, mutations, at the DNA level itself, might be unaccounted for. Nevertheless, this finding is still striking when compared to the degree of biochemical variation currently observed between related taxa in the animal kingdom. In fact, DNA/DNA hybridizations of the single-copy nuclear components of man and chimpanzee have confirmed this similarity [2, 10], thus overcoming the previous methodological objections against electrophoretic analyses. Single-copy DNA hybridizations have the advantage of screening not only the direct composition of most of the active DNA component of each species, regardless of the proteins coded by it, but practically the whole fraction of the genetic material itself that is involved in the process of transcription and translation. On the other hand, physical maps, obtained by endonuclease cleavage of multigene families, such as the 18S-28S ribosomal genes, allowed for the identification of similar gene arrangements between the two species, as evident by the conservation of several restriction sites [6]. Here again, physical maps of the mitochondrial DNA genome [11] have also been relevant in supporting previous findings. The mitochondrial genome evolves some 10 times faster than nuclear DNA, a reason why these studies have been useful in detecting molecular differences between species which might not be present in the nuclear genome. Finally, sequence studies of individual genes, though more limited to highly specific regions of the genome, have also succeeded in demonstrating similarities between man and the chimpanzee [3] due to the striking evolutionary conservation that is evident at the DNA base composition level.

As a cytogeneticist, I have been puzzled by the fact that several human chromosomes have very similar counterparts in the chimpanzee. This is because our chromosome complement, though different from that of the chimpanzee and other great apes, shares several attributes with them. Comparative karyology shows us that the chimpanzee, as all other great apes, has a diploid chromosome number of 48 chromosomes, against 46 in the human. However, karyological similarities between the larger hominoids and man are extensive throughout the whole chromosome complement so that any human chromosome shows a recognizable counterpart in these species and vice versa [8]. Consequently, it is possible to derive the chromosome complement of a species from that of any other by presumed re-arrangements involving relocations of euchromatic regions, such as one fusion and several inversions, and by postulating changes in the organization of heterochromatin and nucleolar organizer regions. It is now know, moreover, that karyological similarities, clearly evident when comparing chromosome number, morphology and banding patterns, are not restricted to a macrostructural level of subcellular complexity.

One of the most intriguing questions was whether similar karyological attributes between species, such as man and the chimpanzee, corresponded to a similarity in gene content and chromosome organization. This was because chromosome bands are morphological attributes resulting from staining procedures not necessarily related to DNA content. G-banding, for example, which is the most widely used method of identifying mammalian chromosomes, might be obtained by the proteolytic digestion of chromosomes by trypsin, an enzyme that does not cleave DNA. When data on comparative gene mapping were unavailable, only partial evidence existed in favour of the positive correlation between chromosome attributes and DNA content, in several mammalian taxa. The mammalian X chromosome of several species was a clear example of a striking evolutionary conservation and was found to contain a common gene cluster across several orders of the mammalian class [12]. In 1969, it was clearly demonstrated that one chromosome trisomy in the chimpanzee was due to the presence of one small acrocentric chromosome, morphologically similar to chromosome 21 in the human [13]. This finding also showed that this trisomy was associated, in the chimpanzee, to a clinical condition similar to that of Down's syndrome in man. It was therefore clear that this chimpanzee chromosome should share similar genes to human chromosome 21 because the same trisomy coexisted with very similar phenotypic effects in the two species.

Two powerful tools soon became available for assigning genes to chromosomes and for carrying out comparative studies between species: (a) in situ hybridization and (b) somatic cell genetics methodologies. The former procedure relies on the detection of molecular hybrids formed between denatured chromosomal DNA and a radioactively labelled DNA (or RNA) molecular probe. This technique, when coupled with chromosome banding, allowed for the detection, at the chromosome band level, of any specific sequence. As chromosome banding, on the other hand, permits the identification of homologous chromosomes between species, interspecific comparisons became possible. The second procedure, i.e. somatic cell genetics, relies on the creation of somatic cell hybrids between a recipient cell line of rodent origin and a donor cell line of another species (e.g. human, chimpanzee). Somatic cell hybrids result from in vitro cell fusion induced by inactivated viruses or polyethylene glycol. These cell hybrids usually conserve one set of chromosomes (the one provided by the recipient cell line) while they loose, or 'segregate', chromosomes of donor origin. As the loss of donor chromosomes is generally random with respect to chromosome number or type, cell hybrid panels might be obtained with different (donor) chromosome combinations. A 90–100% concordance between a donor product (e.g. an enzyme) and a donor chromosome, in a somatic cell hybrid panel, is considered to be indicative of gene assignment. For this to be estimated, donor and recipient products must show some kind of distinctive characteristics, such as different electrophoretic migration so that every cell line in the panel can be screened for a set of biochemical markers. When biochemical and karyological data are compared in a somatic cell hybrid panel, an estimate of concordance is calculated. Concordance is estimated by adding two percentages: (a) the percentage of cases in which one donor marker and one donor chromosome are both present (+/+ class) plus (b) the percentage of cases when the same two markers are absent (–/– class). Conversely, discordance is the percentage of cases in which any one of the two markers is present while the other is absent (+/– plus –/+ classes). Any value of concordance equal or above 90% (corresponding to values of discordance of 10% or less), when statistically significant, is indicative of gene assignment.

Somatic cell genetics and in situ hybridization have been successful in assigning thousands of genes to the human chromosome complement. Conversely, the amount of assigned genes in the chimpanzee is less than 100 [12]. Comparative genetic charts are therefore unequally abundant in data, and much remains to be known before a straightforward comparison might be fully established. However, these studies have been useful in demonstrating that karyological similarities between species are truly indicative of a similar pattern of genome organization.

In fact, comparative gene mapping shows that Primates is an evolutionarily conserved order [12]. A comparison of the human and the chimpanzee gene charts indicates that morphologically similar chromosomes generally contain the same syntenic associations. Thus, the evolutionary conservation of the morphological attributes of human and chimpanzee chromosomes is paralleled by the maintenance of similar gene arrangements in the two species. For example, several genes, which, in the human, form part of the chromosome 12 gene cluster, are also clustered in the corresponding homoeologue chromosome in the chimpanzee. It appears as if our genome were compartmentalized in a similar way, someway split in similar syntenic associations that, moreover, are contained in morphologically similar nuclear structures (chromosomes). In fact, the positive correlation between chromosome morphological attributes and gene content is a general finding in the order Primates, except for some taxa in which chromosome re-arrangement has been prominent, such as the lesser apes. In this group, chromosome shuffling has been so prominent as to preclude the recognition of chromosome homoeologies between lesser apes and man. Not surprisingly, several human gene arrangements appear to be disrupted in the gibbon. Findings of this kind, however, appear to be exceptional, because the evolutionary conservation of syntenic groups is well proven across several orders of the mammalian class [12]. In the domestic cat, for example, a species that belongs to the order Carnivora, several genes of the human chromosome 12 cluster have been found to be associated in a chromosome that is also morphologically similar to human chromosome 12. Thus, several of the genome attributes shared between man, the chimpanzee, other primates and other mammals outside the order Primates appear to be quite old, representing relict arrangements of genes which have remained unchanged for millions of years.

These genetic similarities, however, do not mean that our genome is identically organized. Comparative gene assignment needs yet to increase the density of genetic maps of other mammals, and these studies will eventually lead to a re-evaluation of our present views. When comparing any two species in detail, such as man and the chimpanzee, some minor differences have already been detected. This is the case of chromosome 2 in man, a large biarmed chromosome that is absent in the chimpanzee. In this species, there is a corresponding arm homoeologue to the short arm of human chromosome 2 and another arm homoeologue to the long arm of human chromosome 2. Presumably, a fusion event occurred in the human lineage after the branching off of man from the common stock with the great apes, resulting in a diploid chromosome number of 46 in man (against 48 in the great apes). As expected from cytological data, this presumed fusion must have resulted in an association of two syntenic groups that were separate from one another in the great ape genome. Comparative gene mapping clearly shows that this is the case, though arm homoeologies are apparently the opposite as expected from morphological comparisons. The chimpanzee arm homoeologue to the short arm of human chromosome 2 contains biochemical markers present in the long arm of human chromosome 2 and vice versa. A similar discrepancy is found in the gorilla and the orang-utan when compared to the human, and it might represent an exceptional finding that needs to be further clarified. Moreover, very similar chromosomes, such as human chromosome 20 and its counterpart in the chimpanzee differ with respect to the location of the inosine triphosphatase locus. While in man this marker is in chromosome 20, it has a different location in the chimpanzee, the gorilla and the orang-utan. In these species, inosine triphosphatase is located in the corresponding homoeologue to human chromosome 14, and it is part of a syntenic association similar to that of human chromosome 14+20.

The evolutionary conservation of the gene maps between man, the chimpanzee and other species seems to be rather restricted to structural genes. It does not necessarily apply to the chromosomal distribution of repetitive DNA families despite the fact that some of these families might be in fact very similar, as shown by endonuclease cleavage analyses (physical maps) or sequence studies. The 18S-28S ribosomal genes are a clear example; these human and chimpanzee multifamilies show a remarkable similarity in endonuclease digestion patterns and sequence data [6, 14], but their chromosome distribution differs between these two species. These findings are confirmed in other species of the order Primates in which the distribution of this multigene family appears to be independent of chromosome evolution [9]. In other words, we should not expect these very similar genes to be contained by the expected homoeologous chromosomes, as it happens with structural, single-copy genes. Other studies, involving human homologous satellite DNAs, point to a remarkable conservation, at the molecular level, of some highly repetitive sequences in the chimpanzee and man [5, 7]. Here again, their chromosome distribution has been found to vary between species. Presumably, these highly repetitive sequences of unknown function, though evolutionarily conserved to an extent of showing cross-molecular hybridization in filters and in situ, have been independently amplified in man, the chimpanzee and other great apes. As amplification of a simple sequence motif took place before the phyletic divergence of man from the chimpanzee and other great apes, different amounts of these sequences were amplified in chromosomes that are not necessarily homoeologous. Consequently, these human homologous satellite DNAs do not share a similar chromosome distribution when comparing different species, such as man and the chimpanzee. Here again, as previously commented on for the ribosomal DNA multigenes, differences in chromosome distribution are consequential to the mode and tempo of DNA amplification rather than to differences in sequence content or function [5]. In fact, highly repetitive sequences, such as satellite DNAs, have no known rôle in the mammalian genome, and unlike functional DNA components, they are clearly under less stringent selective pressures.

Genetic similarities between man and chimpanzees are no more striking than the anatomical and behavioural differences that distinguish the two species. Genes and chromosomes are useful tools for establishing taxonomic arrangements and tracing phylogenies. However, they fail in providing a straightforward explanation to the clear contradiction between organic and microstructural evolution. Thus, a clear understanding of what makes the two species so distinctive is still elusive despite our increased knowledge of their chromosome and molecular organization.

References

1 King MC, Wilson AS: Evolution at two levels in human and chimpanzees. Science 1975;188:107–118.
2 Diamond JM: DNA-based phylogenies of the three chimpanzees. Nature 1988;332:685–686.
3 Miyamoto MM, Slightom JL, Goodman M: Phylogenetic relations of humans and African apes from DNA sequences in the 413-globin region. Science 1987;238:969–973.
4 Baldini A, Miller DA, Miller OJ, Ryder OA, Mitchell AR: A chimpanzee-derived chromosome specific alpha satellite DNA sequence conserved between chimpanzees and human. Chromosoma 1991;100:156–161.
5 Gosden JR, Mitchell AR, Seuánez HN, Gosden CF: The distribution of sequences complimentary to satellite I, II and IV DNAs in the chromosomes of the chimpanzee *(Pan troglodytes),* gorilla *(Gorilla gorilla)* and orangutan *(Pongo pygmaeus).* Chromosoma 1977;63:253–271.
6 Arnheim N, Krystal M, Schmickel R, Wilson G, Ryder OA, Zimmer E: Molecular evidence for genetic exchange among ribosomal genes on nonhomologous chromosomes in man and apes. Proc Natl Acad Sci USA 1980;77:7323–7327.
7 Benveniste RE: The Contribution of Retrovirus to the Study of Mammalian Evolution. Molecular Evolutionary Genetics. New York, Plenum Press, 1985.
8 Seuánez HN: The Phylogeny of Human Chromosomes. Heidelberg, Springer, 1979.
9 Seuánez HN: Evolutionary Aspects of Human Chromosomes. Subcellular Biochemistry 10. New York, Plenum Press, 1984.
10 Sibley CG, Alquist JE: The phylogeny of human primates, as indicated by DNA-DNA hybridization. J Mol Evol 1984;20:2–15.
11 Ferris SD, Brown WM, Davidson WS, Wilson AC: Extensive polymorphism in the mitochondrial DNA of apes. Proc Natl Acad Sci USA 1981;78:6319–6323.
12 O'Brien SJ, Seuánez HN, Womack JE: Mammalian genome organization: An evolutionary view. Annu Rev Genet 1988;22:323–351.
13 McClure HM, Belden KG, Pieper WA: Autosomal trisomy in a chimpanzee: Resemblance to Down's syndrome. Science 1969;165:1010–1011.
14 Wilson GN, Knoller M, Szura L, Schmickel RD: Individual and evolutionary variation of primate ribosomal DNA transcription initiation regions. Mol Biol Evol 1984;1:221–237.

Héctor N. Seuánez, MD, PhD, Instituto Nacional do Cancer, Genetics Section,
Universidade Federal do Rio de Janeiro, Praça Cruz Vermelha 23, 6° andar, 20.330,
Rio de Janeiro (Brazil)

Eder G, Kaiser E, King FA (eds): The Role of the Chimpanzee in Research. Symp, Vienna 1992. Basel, Karger, 1994, pp 43–55

Similarities and Differences in the Neonatal Behavior of Chimpanzee and Human Infants

Kim A. Bard

Yerkes Regional Primate Research Center, Emory University, Atlanta, Ga., USA

Introduction

The behavioral research project that is presented in this article was designed to describe the neurobehavioral integrity, behavioral organization and regulatory ability of nursery-reared chimpanzees during their first month of life. This study draws heavily upon the human literature of neonatal behavior which has documented the skills that infants bring into the world. Based on the last 20 years of infancy research, we now understand that the human newborn can actively participate in early social interactions and can actively regulate behavioral state. The social competence of the newborn can readily be seen in the face-to-face context [1], where skills include alerting, brightening of the eyes and face, following sights and sounds, moving in synchrony with speech rhythms and matching some facial expressions [2, 3]. Regulatory abilities are evident, for instance when stimulation is excessive and the neonate turns away and self-consoles.

The relative amount of information on normative human behavior stands in sharp contrast to the extent of knowledge concerning neonatal chimpanzees [4, 5]. Newborn chimpanzees are in constant physical contact with their mothers, and new mothers are often shy of observers. Motoric reflexes, like rooting and grasping, are present at birth, but chimpanzees are unable to remain clinging to their mother without her active support. Chimpanzee mothers engage in face-to-face interactions with their infant [6]. We wondered to what extent there are similarities between chimpanzees and humans, either in early social competence or early regulatory ability, similarities that could be accounted for by common evolutionary history.

This study began using human neonatal behavior as a guide. We searched for aspects of behavior that characterized each species. The specific goals were (1) to describe the neurobehavioral integrity of neonatal chimpanzees, (2) to describe development during the neonatal period and (3) to compare chimpanzee neonates with human neonates. The similarities between newborn chimpanzees and newborn humans are really quite remarkable.

Methods

Subjects

Thirteen nursery-reared chimpanzees and 42 human infants were the subjects of this study. The chimpanzees were raised in the Great Ape Nursery of the Yerkes Regional Primate Research Center of Emory University. At Yerkes, there are many chimpanzees who have sufficient maternal behaviors and, therefore, are able to raise their own offspring. Unfortunately, despite our best efforts, at Yerkes there are also chimpanzees who do *not* have sufficient maternal behaviors to care for their infants. The basic mothering skills that are lacking in these mothers include simply picking up the baby [6]. When extensive observations indicate that a particular mother is inadequate in the care of her newborn, then the infant is placed in our nursery. This was the case for the 13 subjects of this study. Chimpanzees were tested every other day from 2 days after birth to 30 days of age. Human infants were studied in Dr. Barry Lester's laboratory at Brown University and Bradley Hospital, in Rhode Island. The human infants were born to middle-to-upper class mothers with no history of drug use throughout the pregnancy. Human infants were tested at day 2 and at day 30.

Procedure

Chimpanzee and human neonates were studied with an assessment protocol developed by Brazelton [7] for use with human neonates. The test, called the Brazelton Neonatal Behavioral Assessment Scale or the NBAS for short, is based on a standard neurological examination. However, the 20–30-min NBAS examination also includes items that assess the newborn's social capacities, motor development and regulatory abilities. The NBAS was administered to chimpanzee neonates without modification. Since chimpanzees were raised with humans as their primary caretakers, it was appropriate to assess their responses to human-based stimuli.

Conceptually, the NBAS is divided into 7 areas or clusters of behavior [8]. Data concerning both human and chimpanzee performance in 5 of these clusters are presented in this article. Specifically, these clusters of behavior are orientation, motor performance, autonomic nervous system stability, range of state and state regulation. The manner in which chimpanzees typically performed in the NBAS is illustrated with individual items within each of these clusters of behaviors.

A repeated measures analysis of variance was conducted on each of the 28 items of the NBAS and on scores in each of the 5 clusters. Chimpanzee infant performance was compared with human performance. In addition, a developmental question was addressed by comparing behavior in the beginning of the neonatal period (that is on day 2) with behavior at the end of the neonatal period (specifically on day 30).

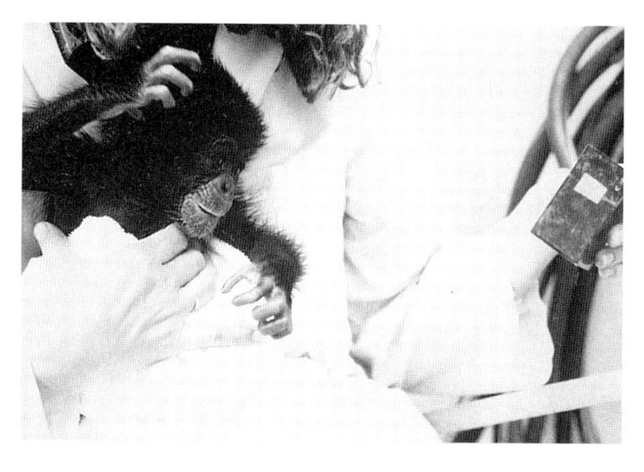

Fig. 1. Neonatal chimpanzees, such as Carole at 18 days of age, turn and locate the source of inanimate sounds.

Results

Orientation

The 7 items of the orientation cluster assess the neonates ability to follow and locate sights and sounds. The orientation cluster includes 3 nonsocial items (a visual item, an auditory item and a visual/auditory combined stimulus), 3 social items (again a visual item, an auditory item and a visual/auditory combined stimulus) and an overall measure of quality of alertness.

Nonsocial stimuli include a red ball and a red rattle. Carole, at 18 days of age, turns to the sound of the rattle and visually locates it (fig. 1). Social stimuli include the examiners' face and voice. Debbie, at 19 days of age, turns to the sound of my voice and locates my face (fig. 2). Both human sounds and chimpanzee species-typical grooming sounds have been used as social stimuli, and neonatal chimpanzees responded equally well to both types of social sound [9].

The mean scores for human and chimpanzee neonates on the item 'orientation to an animate auditory and visual stimulus', are shown in figure 3. As a point of reference, the scoring system for each orientation item of the NBAS was designed in such a way that the average human would score at the midpoint of the scale. In other words, the mean score for humans at day 2 was planned to be a 5. So, this figure shows how well newborns oriented to the examiners' animated and talking face. There was no difference between chimpanzee and human performance on this social item. Both species, however, did show significant improvement in performance from day 2 to day 30. In general, we

Fig. 2. Neonatal chimpanzees, such as Debbie at 19 days of age, turn and locate the source of social sounds.

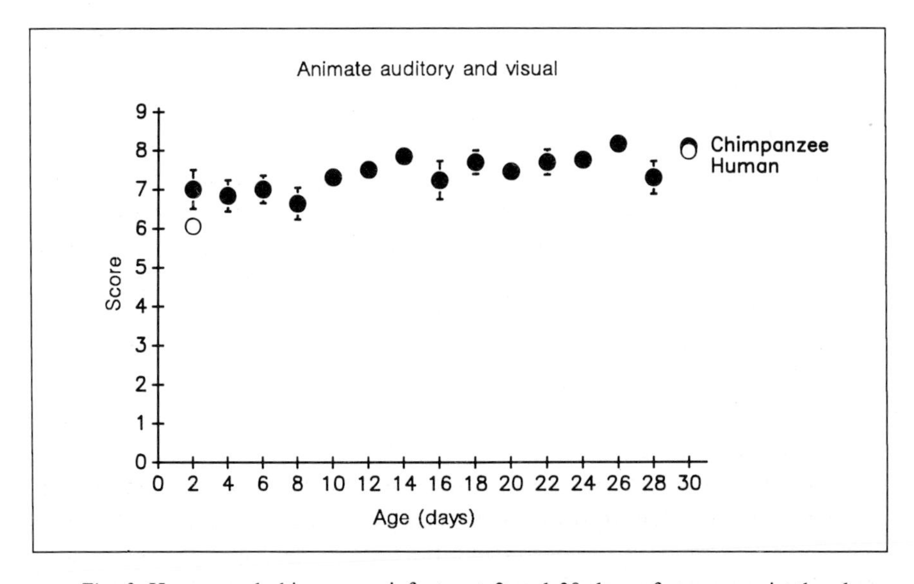

Fig. 3. Human and chimpanzee infants, at 2 and 30 days of age, score in the above average range for orientation to the examiner's talking and animated face.

found that performance is better and more consistent in response to the social stimuli compared with the nonsocial stimuli. This was true for chimpanzee as well as human neonates [9].

An additional score is given for the quality and duration of the best periods of alertness throughout the examination. A significant difference was found

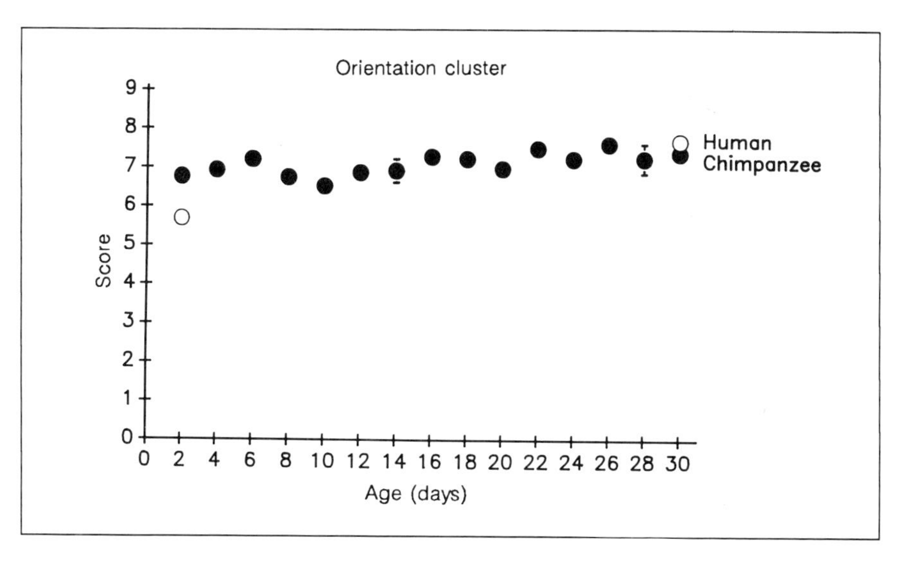

Fig. 4. Chimpanzee and human neonates performed equally well on the orientation cluster of behaviors.

between the species on this item. ($F_{1,43} = 20.78$, $p < 0.001$). Neonatal chimpanzees, from 2 days of age, alert quickly and stay alert for long periods, with consistent quality. In comparison, 3-day-old human infants are alert for periods of only 'moderate' duration, there may be a delay before they respond and they may be uneven in the quality of their responsiveness. Developmental changes were found in alertness: by 30 days of age, both human and chimpanzee neonates exhibited a significant improvement in the quality of alertness.

Scores on all 7 orientation items were averaged together to form the orientation cluster score for each subject. Figure 4 shows the mean orientation cluster scores for all subjects. The orientation performance of chimpanzees and humans, on day 2 and day 30, was strikingly similar. In fact, there was no statistical difference between the species. There was a significant improvement in orientation performance for both species from day 2 to day 30. Both human and chimpanzee neonates have the capacity for sustained attention, to visual and auditory stimuli, of both a social and nonsocial nature.

Motor Performance

The 5 items of the motor cluster assess the neonate's muscle control and coordination, general tone of the musculature and motor activity. Specific maneuvers include defensive reaction and pull-to-sit. No significant difference was found between the performance of chimpanzee and human neonates in their defensive reaction. In the pull-to-sit maneuver the examiner takes the infant by the hands and smoothly pulls to a sitting position. The important aspect of this item is how well the baby can hold his head without support. Chimpanzees have excellent head control even at birth. Chimpanzees performed at the top of the scale which indicates that there was no head lag as the baby was pulled to a sitting position, and, once sitting, chimpanzees held their head upright for a full minute. Chimpanzees performed significantly better than human infants on the pull-to-sit item ($F_{1,41} = 25.64$, $p < 0 .001$).

Scores on the 5 motor items were averaged together to form the motor performance cluster score. When performances on all the motor items were averaged together, in the motor cluster, newborn chimpanzees do *not* significantly differ from newborn humans in their overall motor performance. Developmental changes, however, were found. In both humans and chimpanzees, motor performance improved significantly from day 2 to day 30 ($F_{1,42} = 16.96$, $p < 0 .001$).

Autonomic Stability

The autonomic nervous system stability cluster measures signs of stress in the autonomic nervous system; specifically, it measures the number of tremors, the number of startles and changes in skin color. Skin color, however, was not used in this study because in any dark-skinned baby, chimpanzee or human, changes in skin color are difficult to see. There were no significant differences between human and chimpanzee newborns in the number of startles, the amount of tremors or the autonomic stability cluster scores. In addition, there was no significant developmental change from day 2 to day 30.

Range of State

Range of state concerns the neonates' characteristic state of arousal during the examination. Changes in behavioral state are measured throughout the 30-min examination. Arousal is measured by noting the time and intensity of behavioral state changes. In addition, the number of items to which the infant reacts with irritable fussiness and the total number of state changes are noted. The examination begins with the infant asleep. Chimpanzee neonates look very much like human neonates when they are sleeping, exhibiting both deep sleep and REM sleep (fig. 5). One goal of the NBAS examination is to observe the infant's responses while in *different* behavioral states. Chimpanzees *may* reach a highly aroused state such as crying. Chimpanzee neonates tend to maintain a quiet awake state throughout the majority of the examination.

The small but significant difference between chimpanzee and human neonates in the range of state cluster scores ($F_{1,42} = 9.15$, $p < 0.005$) is illustrated in figure 6. Human infants scored higher than neonatal chimpanzees. A

Fig. 5. Chimpanzee neonates exhibit both deep and REM sleep.

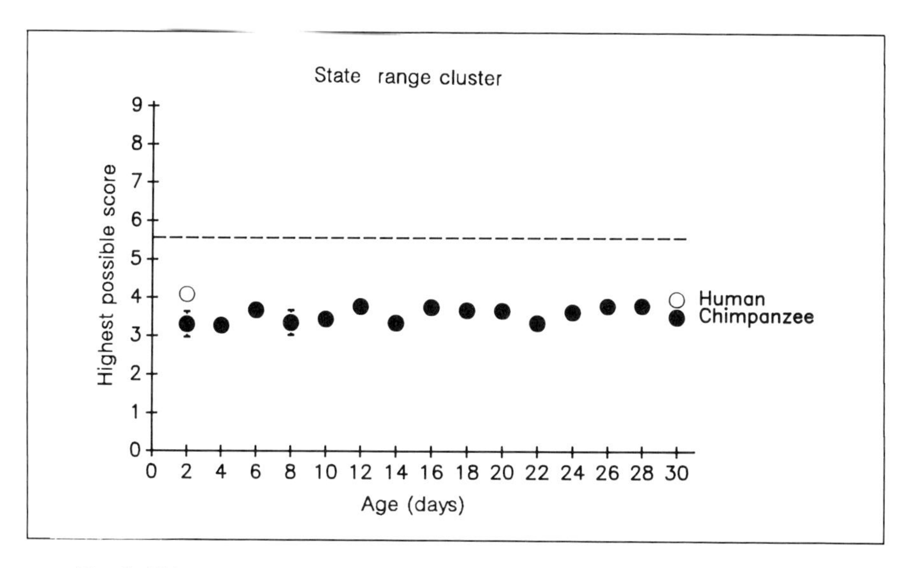

Fig. 6. Chimpanzee neonates scored significantly lower than human neonates on the range of state cluster of behaviors, reflecting the chimpanzees' lower general arousal.

higher score on this cluster reflects both a higher intensity of arousal and arousal which occurs earlier in the examination. The chimpanzees' *lower* score reflects the fact that chimpanzees rarely maintain a crying state and, if chimpanzees cry, it is only at the end of the examination, with the presentation of the most intrusive items.

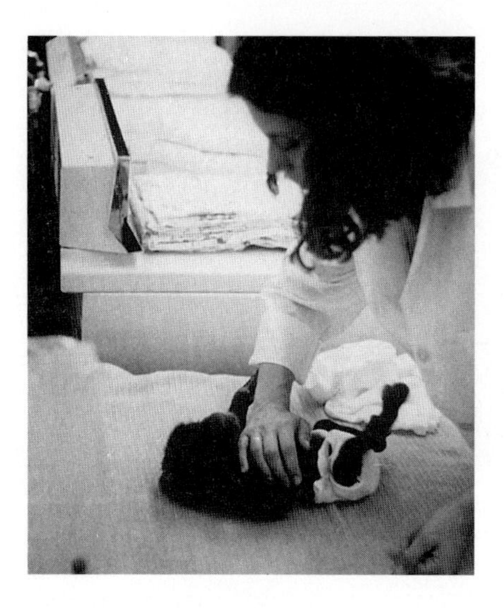

Fig. 7. When the neonate cries, consolability is measured by the amount of examiner intervention. Carole, who is 4 days old, quiets when the examiner places her hand on Carole's belly.

State Regulation

The items in the state regulation cluster detail the neonate's success in reducing arousal and the amount of help needed from the examiner for the infant to return to a quiet awake state. This cluster concerns the mechanisms which underlie the neonates' change of state. Higher scores in state regulation indicate higher scores in one or more of the following 4 items: consolability, self-quieting, hand-to-mouth behavior and cuddliness. The performance of chimpanzee and human neonates on each of these items is considered in the following paragraphs.

Consolability is measured by the number of maneuvers used to bring the infant's level of arousal from a fussy or crying state down to a quiet awake state. Figure 7 shows Carole, at 4 days of age, quieting with the examiners' hand on her belly. Hand-on-the-belly is at the low end of a series of 8 consoling procedures. There was *not* a significant difference between the species on consolability.

Active self-quieting behaviors can be observed in human and chimpanzee neonates. The types of self-quieting activity include focusing attention, either on a visual or auditory stimulus, sucking on a fist or thumb and maintaining a preferred body posture. Chimpanzee and human neonates exhibit a well-developed ability to self-quiet in the neonatal period. There were, however, large individual differences in neonates' ability to self-quiet. No significant

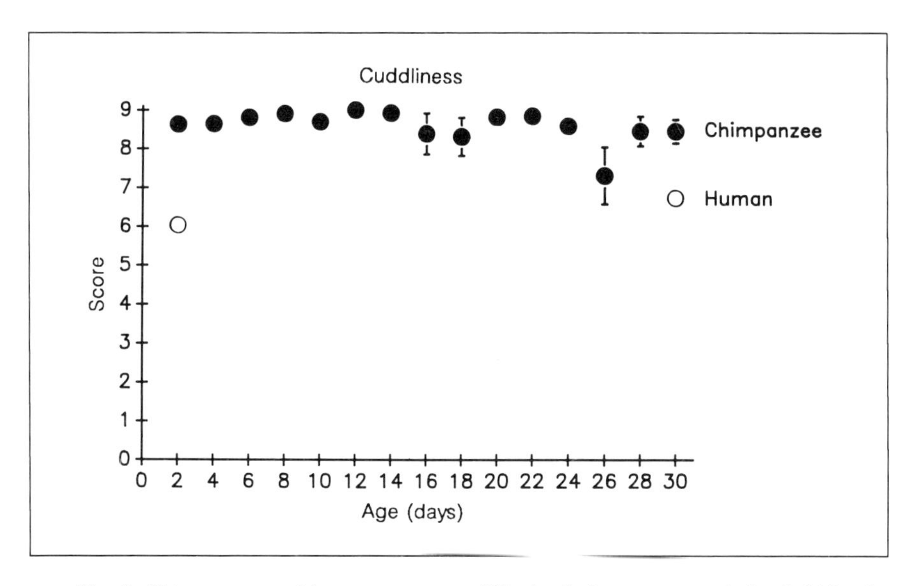

Fig. 8. Chimpanzee and human neonates differ in their response to being held by the examiner.

differences were found between the species. Hand-to-mouth ability reflects any attempts to bring the hand, fist, wrist or fingers to the mouth. Some newborn chimpanzees, as well as some humans, are able to suck their thumb [10] but no significant differences were found between chimpanzee and human newborns in their hand-to-mouth ability.

Cuddliness is a measure of the infant's response to being held. Neonatal chimpanzees almost always quiet when they are picked up and held. In addition to molding and relaxing, chimpanzees grasp and cling to the examiner, thus earning a top score on the cuddliness scale (fig. 8). Out of the 4 items of the state regulation cluster, species differences were found only in this item, namely the measure of cuddliness ($F_{1,42} = 24.85$, $p < 0.001$).

Scores on these 4 items were averaged together to arrive at a state regulation cluster score (fig. 9). Neonatal chimpanzees scored significantly higher than neonatal humans, indicating the superior ability of chimpanzees to regulate their own behavioral state ($F_{1,41} = 11.08$, $p < 0.002$).

The last of the 28 items of the NBAS is not averaged into any cluster but, for the sake of completeness, it is important to mention: number of smiles. Chimpanzees often smiled while engaged in physical maneuvers, such as supporting their own weight. In addition, neonatal chimpanzees often smiled during orientation items (fig. 10). Carole, at 15 days of age, smiles as she follows the examiner's animated and talking face.

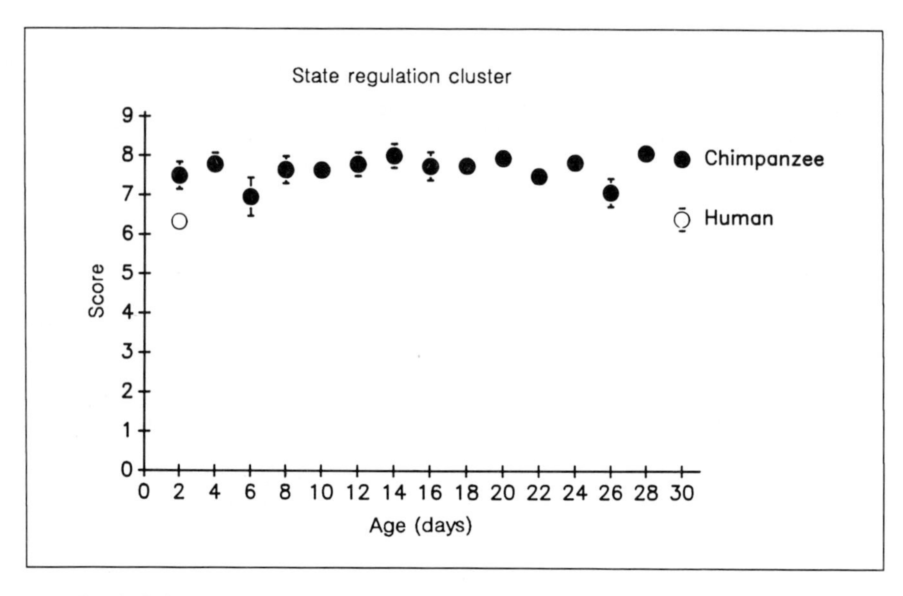

Fig. 9. Chimpanzee neonates scored higher than human neonates on the state regulation cluster of behaviors.

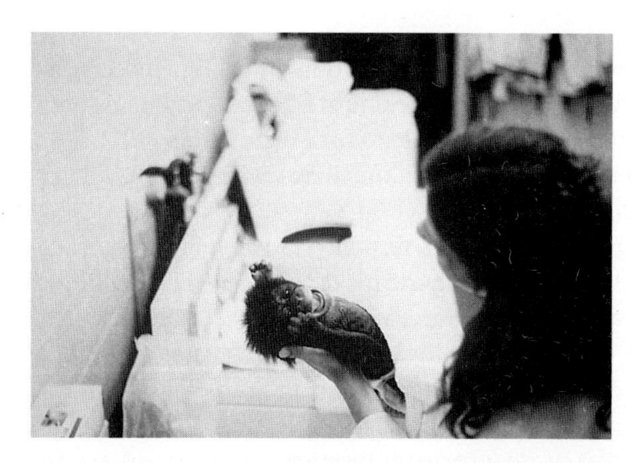

Fig. 10. Carole, a 15-day old chimpanzee neonate, smiles in response to social orientation items.

The average number of smiles during the NBAS examination for chimpanzee and human neonates is illustrated in figure 11. There was no significant difference between the species. The large standard error bars indicate that there is a great deal of individual variation for chimpanzee subjects. Significant developmental changes were found: both chimpanzee and human newborns smiled more at 30 days of age than they did at 2 days of age.

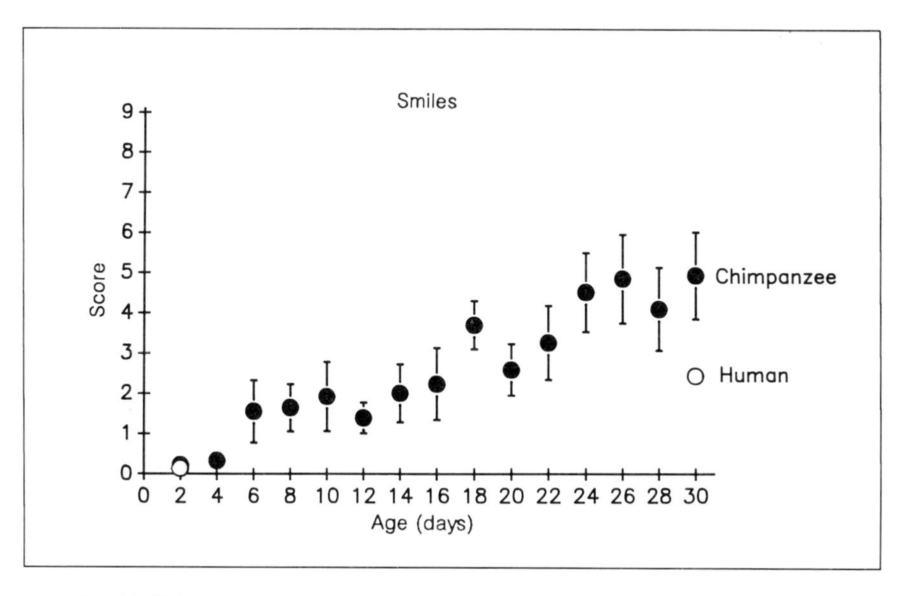

Fig. 11. Chimpanzee and human neonates smiled equally often during NBAS examinations.

Discussion

In summary, the NBAS examination proved extremely useful. With the NBAS examination, the neurobehavioral integrity of chimpanzees can be described [9, 11–13]. Moreover, by using the same procedure with each group of neonates, direct comparisons can be made between the performance of chimpanzee and human newborns, between chimpanzee groups raised in different environments [14] and between human groups raised in different cultures [15].

This research was begun 5 years ago. The goal was to develop a temperamental assessment for chimpanzees that paralleled the one developed for use with rhesus monkeys [16]. After trying unsuccessfully to apply the rhesus monkey version to chimpanzees, I learned the original human protocol; the rhesus monkey and human protocols *are* substantially different. The human protocol could be used, unchanged, with chimpanzees and, in fact, the rhesus monkey protocol could *not* be used with chimpanzees. This highlights the initial impression, which is now supported by the data, that chimpanzee behavior is much more similar to that of human behavior. Chimpanzee behavior, especially in the neonatal period, is very different from that of rhesus monkeys.

The first conclusion is that chimpanzee and human neonates are more similar than different. This conclusion is based first on the fact that the *same* assessment protocol is used successfully with both species. There are striking similarities in the behavior of chimpanzees and humans during the neonatal period. In orientation, chimpanzees attend to the same stimuli as do human infants with a similar quality of engagement. We view these striking similarities in early orientation ability as a challenge to the notion that humans have a unique propensity to engage in face-to-face communicatory and social interactions. Over all the motor performance items, chimpanzee performance is nearly identical to that of human infants. Overall, the amount of startles and tremors reflecting stress in the autonomic nervous system is comparable for both species. Therefore, in 3 out of 5 clusters of behavior, and in the majority of the 28 items of the NBAS, no significant differences are found between newborn chimpanzees and humans.

Behavioral state is one of the most important considerations when assessing newborn capacities. The two NBAS clusters related to behavioral state revealed significant differences between the two species. In contrast to human infants, chimpanzees maintain a quiet alert state throughout the examination. Chimpanzee neonates utilize their own behaviors, or those of the examiner, to regulate their state to a greater extent than do human neonates. Differences in behavioral state account for most of the differences between chimpanzee and human neonates.

Although species differences exist, they are of a *quantitative*, rather than a qualitative nature. This study represents a documentation of the commonalities, as well as the differences, among primate species in the area of neonatal behavior. The long term goal is the creation of a normative data base. Thus, a prospective and longitudinal study of individual differences in terms of temperament, neurological functioning, and risk status can be conducted. We hope that this will lead to a greater understanding of developmental processes that culminate in adult competence.

Chimpanzees are interesting to study for their own sake. In addition, by understanding chimpanzees, we can expand our knowledge of our own primate heritage. Studies of neonatal behavior serve to illustrate commonalities and unique characteristics of each primate species. Knowledge of the evolutionary roots of human behavior provides an important adaptive context, one that enriches our appreciation of all primate behavior.

Acknowledgements

This work was supported by NIH grants RR-00165, RR-03591 and RR-06158 to the Yerkes Regional Primate Research Center of Emory University, NICHD Intramural Program Funds to the Laboratory of Comparative Ethology, Director Dr. Stephen J. Suomi, and NRSA HD-07105. Special acknowledgment is given to Dr. Frederick A. King, Ms. Carolyn L. Fort, Dr. Kathleen A. Platzman, Dr. Barry M. Lester, Dr. Stephen J. Suomi, and Mr. Josh Schneider for assistance with graphics. Permission to reprint figures was graciously provided by *The Newsletter*, published by the Primate Foundation of Arizona, directed by Ms. Jo Fritz.

References

1 Brazelton TB: Evidence of communication in neonatal behavioral assessment; in Bullowa M (ed): Before Speech: The Beginning of Interpersonal Communication. New York, Cambridge University Press, 1979.
2 Meltzoff AN, Moore MH: Imitation of facial and manual gestures by human neonates. Science 1977;198:75–78.
3 Kugiumutzakis G: Intersubjective vocal imitation in early mother-infant interaction; in Nadel J, Camaioni L (eds): New Perspectives in Early Communication Development. London, Routledge, 1992.
4 Plooij FX: The Behavioral Development of Free-Living Chimpanzee Babies and Infants. Monographs on Infancy. Norwood, Ablex, 1984, vol 3.
5 Goodall J: The Chimpanzees of Gombe: Patterns of Behavior. Cambridge, Belknap Press, 1986.
6 Bard KA: Evolutionary roots of intuitive parenting: Maternal competence in chimpanzees. Early Dev Parenting, in press.
7 Brazelton TB: Neonatal Behavioral Assessment Scale, ed 2. Clinics in Developmental Medicine No 88. Philadelphia, Lippincott, 1984.
8 Lester BM: Data analysis and prediction; in Brazelton TB (ed): Neonatal Behavioral Assessment Scale. Clinics in Developmental Medicine. Spastics International Medical Publications. Philadelphia, Lippincott, 1984, pp 85–96.
9 Bard KA, Platzman KA, Lester BM, Suomi SJ: Orientation to social and nonsocial stimuli in neonatal chimpanzees and humans. Infant Behav Dev 1992;15:43–56.
10 Bard KA, Hopkins WD, Fort C: Lateral bias in chimpanzees *(Pan troglodytes)*. J Comp Psychol 1990;104:309–321.
11 Bard KA, Platzman KA, Suomi SJ, Fort CL, Lester BM: Neurobehavioral assessment in neonatal chimpanzees and humans: Motor performance, range of behavioral state, state regulation, and autonomic nervous system stability. Behav Neurosci, submitted.
12 Hallock MB, Worobey J, Self PA: Behavioral development in chimpanzees *(Pan troglodytes)* and human newborns across the first month of life. Int J Behav Dev 1989;12:527–540.
13 Redshaw M: A comparison of neonatal behavior and reflexes in the great apes. J Hum Evol 1989; 18:191–200.
14 Bard KA: Very early social learning: The effect of neonatal environment on chimpanzees' social responsiveness. Proc Cong IPS, in press.
15 Nugent JK, Lester BM, Brazelton TB: The Cultural Context of Infancy. Norwood, Ablex, 1989, vol 1: Biology, culture, and infant development.
16 Schneider ML, Moore CF, Suomi SJ, Champoux M: Laboratory assessment of temperament and environmental enrichment in rhesus monkey infants *(Macaca mulatta)*. Am J Primatol 1991; 25:137–155.

Kim A. Bard, PhD, Division of Reproductive Biology, Yerkes Regional Primate Research Center, Emory University, Atlanta, GA 30322 (USA)

Eder G, Kaiser E, King FA (eds): The Role of the Chimpanzee in Research.
Symp, Vienna 1992. Basel, Karger, 1994, pp 56–67

The Chimpanzee: A Useful Model for the Human in Research on Hormonal Contraception

Ronald D. Nadler

Yerkes Regional Primate Research Center, Emory University, Atlanta, Ga., USA

Rationale

Use of the chimpanzee as a model for the human in research on hormonal contraception is based on evidence which suggests that (1) an animal model is required to address several unresolved questions regarding the use of oral contraceptives (OCs) by women and (2) the chimpanzee is an especially appropriate substitute for the human in those particular areas for which the action of OCs is unclear. A major area of interest in which the influence of OCs in women is controversial is that of sexual behavior, both with respect to the woman who uses the contraceptive and to her male partner. Although early reports on the sexual behavior of women using OCs suggested generally positive effects on libido and sexual behavior, subsequent data have been mixed [1–3]. Although most women report little or no apparent effects of the OC, some report adverse effects on both libido and sexual behavior [4]. Interpretation of adverse effects in these respects is complicated, moreover, by hormonally induced changes in the genital tissues of some women who use OCs [5].

The difficulty in assessing the basis of hormonally mediated effects on behavior in women is related to the relatively low behavioral responsiveness of women to hormonal influences and their relatively high sensitivity to psychological, social and cultural variables [6, 7]. It is for these reasons that an animal model is useful for clarifying possible behavioral influences of OCs which may be primarily of hormonal origin. Early laboratory research on the chimpanzee suggested that this species could serve as a useful substitute for the human in issues related to sexual behavior for several reasons [8–10]. Most relevant in the present context is the partial emancipation of the chimpanzee's sexual behavior from hormonal control, reflected by mating during the menstrual cycle which is temporally dissociated from the midcycle phase when endogenous concentrations of estrogen are elevated (and progesterone concentrations are minimal). Although the chimpanzees in these laboratory studies mated primarily during the midcycle phase, they also mated at other phases of the cycle, unrelated to elevated estrogen concentrations. The chimpanzee, thereby, demonstrated some degree of behavioral independence from hormonal control, similar to the human, while clearly retaining a greater degree of responsiveness to hormonal influences than is characteristic of humans.

The cyclic pattern of changes in sex hormone concentrations during the menstrual cycle is another relevant area in which the chimpanzee is remarkably similar to the human, more similar than the simian primates [11, 12]. The chimpanzee resembles the human, for example, in the midluteal phase elevation in estradiol, a feature that is minimal or absent in the simians. The female chimpanzee also possesses a highly specialized anogenital swelling which is exquisitely sensitive to changing concentrations of estrogen and progesterone [11, 13, 14]. Although the anogenital swelling is generally valued by researchers as an external index for monitoring progress of the natural menstrual cycle in the chimpanzee, it is useful in research on OCs because the genital component of the swelling is homologous to the labia minora of women. Both the woman's and the female chimpanzee's labial tissues respond to hormonal changes, but the tissues of the chimpanzee are more sensitive to such changes and/or are more conspicuous in their response.

The chimpanzee is a useful model for investigations on OCs, therefore, because of a combination of relevant similarities and advantageous differences between it and the human. On the one hand, the chimpanzee resembles the human with respect to (1) the regulation of its sexual behavior, (2) the cyclic pattern of sex hormone concentrations during the menstrual cycle of the female and (3) the responsiveness of the female's genital tissues to hormones. On the other hand, the behavioral and genital responses of the chimpanzee to fluctuations in hormone concentrations per se are clearer than those of the human, i.e. the chimpanzee responds more clearly to hormonal fluctuations of the type that accompany the use of OCs. The chimpanzee may be viewed as a sensitive index to effects of OCs that may have relevance to the human but that are generally obscured in the human by complex cognitive processes which are less developed in the chimpanzee.

Hypotheses

Several hypotheses were proposed regarding the action of an OC in the chimpanzee [15, 16], based, in part, on the well-documented reduction in endogenous sex hormone concentrations that results from the use of OCs by women [17–19]. It was hypothesized that the administration of a combined (estrogen/progestin) OC to the female chimpanzee would produce:

(1) relatively low and tonic levels of endogenous ovarian steroids; this hypothesis follows by extrapolation from the results in women;

(2) relatively low and tonic levels of female anogenital swelling, related to the dosage of estrogen in the combined OC; this hypothesis follows from previous research on the hormonal regulation of anogenital swelling in the chimpanzee [11];

(3) relatively low female attractivity and proceptivity, but without effect on receptivity [20], (attractivity is measured by male sexual initiative, proceptivity is measured by appetitive female sexual behavior and receptivity is measured by the female's response to male sexual initiative); this hypothesis follows from previous research on the hormonal regulation of sexual behavior in nonhuman primates, including the role of the anogenital swelling in sexual activity of the chimpanzee [21, 22];

(4) relatively low rates of copulation during the menstrual cycle; this hypothesis follows from hypothesis 3 above;

(5) the clearest reduction in copulation in laboratory pair tests of the type in which the female controls sexual access, in comparison with the traditional laboratory pair test; this hypothesis follows from previous research which demonstrated that the female chimpanzee's ability to regulate sexual interactions in the traditional laboratory pair test may be compromised by a dominant male [8, 9].

Experimental Design

The subjects were 9 pairs of chimpanzees composed of 9 females and 4 males. The treatments were (a) a combined OC consisting of ethinyl estradiol (EE_2; in dosages of 50, 100 or 400 μg) plus 0.5 mg norethindrone and (b) a placebo. Each treatment, OC and placebo, was administered (in juice or food) for 4 cycles, 2 cycles without behavioral testing followed by two cycles with testing, balanced for order of treatment. The 2 cycles without testing were used (1) to confirm normal cyclicity prior to testing with the placebo and (2) to allow hormone levels to stabilize prior to testing with the OC. Each dosage of EE_2 and placebo was administered to 3 females. Two types of behavioral tests were conducted during each treatment, the traditional laboratory pair test or free-access test (FAT) and the restricted-access test (RAT) with female control of access. The FAT was conducted in a single cage with the male and female freely accessible to each other. The RAT was initiated with the male and female in separate cages, with an interconnecting door that was opened only when the female performed a lever-pressing task.

Analyses

Data analysis and hypothesis testing were conducted on (1) endogenous and exogenous sex hormone concentrations throughout the natural and OC menstrual cycles, (2) anogenital swelling scores throughout the natural and OC menstrual cycles, and (3) sexual and social behavior scores throughout the natural and OC menstrual cycles.

Results

Endogenous and Exogenous Hormone Concentrations in the Female

During the natural cycle with placebo, cyclic patterns of sex hormones typical of the chimpanzee were recorded. As a result of OC administration, cyclicity in endogenous hormone concentrations was abolished, and relatively stable concentrations of the exogenous hormones were established which, in the case of EE_2, were related to the dosage of EE_2 administered. Serum concentrations of luteinizing hormone, progesterone and testosterone were maintained at concentrations typical of the early to midfollicular phase regardless of the dosage of EE_2 administered, whereas estradiol decreased with the increasing dosages of EE_2. Serum concentrations of norethindrone were low but measurable, and EE_2 increased with the increasing dosages. The decrease in serum estradiol concentrations and the reciprocal increase in EE_2 concentrations resulted in an approximately balanced exchange of the exogenous estrogen for the endogenous one. Urinary metabolites of the endogenous and exogenous hormones exhibited patterns that were comparable to those in serum.

Female Anogenital Swelling

During the natural cycle with placebo, the anogenital tissues exhibited the cyclic pattern typical of female chimpanzees, including an extended midcycle phase with maximal anogenital scores of 4. The cyclic increase in anogenital swelling was abolished by OC administration and the degree of relatively stable swelling which was induced was related to the dosage of EE_2 administered. The mean anogenital score with the 50-μg dosage of EE_2 was 1.3 ± 0.1, increased to 1.5 ± 0.1 with the 100-μg dosage and increased further to 2.5 ± 0.1 with the 400-μg dosage. The overall differences among the scores were statistically significant in an analysis of variance, as were tests of orthogonal comparisons.

Table 1. Percentage of test days with copulation (cycle rate) by chimpanzees during natural menstrual cycles with a placebo (PL) and during cycles with an OC, in two types of pair test, the FAT and the RAT with female control of access

	FAT		RAT		Mean rate	
	PL	OC	PL	OC	PL	OC
$Anna_{50}$ × Harvey	12	0	26	0	19	0
$Flora_{50}$ × Phineas	4	0	0	0	2	0
$Wenka_{50}$ × Frans	93	90	9	40	51	65
$Sonia_{100}$ × Harvey	34	0	12	15	23	7
Gay_{100} × Rogger	23	0	0	0	11	0
Li'l One_{100} × Phineas	45	40	44	4	45	22
$Cheri_{400}$ × Frans	0	0	0[1]	0	0	0
$Jacqueline_{400}$ × Rogger	4	0	0[1]	0	2	0
$Barbi_{400}$ × Harvey	31	9	18	0	25	5

A subscript after the female's name indicates the daily dosage of EE_2 (μg), combined with 0.5 mg norethindrone, the female received in the OC cycle.

[1] The female did not lever-press to allow sexual access on any test days during the cycle.

FATs of Sexual Behavior

During the natural cycle with placebo, 8 of the 9 pairs copulated during 31% of the test days on average; the percentage of test days with copulation ranged from 4 to 93% for the 8 pairs (table 1).

One pair ($Cheri_{400}$ × Frans) failed to copulate on any of the tests of either type, whether with placebo or OC. All 8 pairs that copulated during the natural cycle copulated on a lower percentage of test days during the OC cycle – a significant result. The difference in copulatory rates, however, was directly related to the rate of copulation in the natural cycle. The pairs that copulated at the higher rates in the natural cycle also copulated in the OC cycle and did so at very nearly the same rates. Those pairs that copulated at low rates in the natural cycle, on the other hand, failed to copulate at all in the OC cycle. There was no relationship between copulation and the dosage of EE_2 the females received; 1 female copulated during the OC cycle at each dosage.

The rate of male sexual initiative per test day in the OC cycle was approximately half the rate in the natural cycle (0.8 ± 0.3 vs. 1.7 ± 0.05, $p < 0.025$), and the rate of unsolicited proceptive behavior was lower in all the females that exhibited the behavior (0.04 ± 0.04 vs. 0.4 ± 0.2, OC < placebo for all n). In fact, only 1 female showed unsolicited proceptive behavior in an OC cycle. The females' receptivity, on the other hand, in terms of their responsiveness to male copulatory attempts and male solicitations, was approximately the same regardless of treatment.

RATs of Sexual Behavior

The results of the RATs, contrary to our hypothesis, were not as clear as those of the FATs, related in part to an absence of copulation during the natural cycle in 4 of the pairs, i.e. there was no possibility of an adverse effect of the OC on copulation in 4 of the 9 pairs (table 1). Of the remaining 5 pairs, 3 exhibited less copulation in the OC cycle, but 2 showed some increase. As in the case of the FATs, however, male sexual initiative was clearly lower in the OC cycle (0.7 ± 0.3 vs. 1.7 ± 0.3, $p < 0.01$), as was unsolicited proceptive behavior (0.2 ± 0.1 vs. 0.7 ± 0.2, $p < 0.01$, OC < placebo for all n). Also in agreement with the data from the FATs, female responsiveness to male initiative was not influenced by OC administration.

Sexual Behavior in FATs and RATs Combined

Overall, 7 of the 8 pairs that copulated during the natural cycles copulated at lower rates during the OC cycles (table 1). The only pair that did not show this overall effect of the OC ($Wenka_{50}$ × Frans) was unusual in several ways. The copulation of this pair bore no relationship to the female's pattern of anogenital swelling during the natural cycles, suggesting that the behavior of this pair was unrelated to the female's hormonal condition generally. The female of this pair, moreover, did not present sexually to the male with her ventrum close to the ground; the female did not present in the manner most indicative of high sexual arousal [9].

Sexual Behavior and Social Interaction

In order to gain some insight into the basis for differences in sexual behavior among the pairs and for differences among them in response to the OC, we examined the relationship between their sexual activity and their social interactions (table 2). All 4 pairs in which the female's dominance vis-a-vis the male was moderate or high failed to copulate during both of the OC cycles. Four of the remaining 5 pairs in which the female's dominance was low copulated to some extent during the OC cycles. The only pair with low female dominance which did not copulate during either of the OC cycles (Gay_{100} × Rogger) was the only pair in which there was no grooming, i.e. no affiliative interactions at all.

Discussion

The combined OC had adverse effects both on the anogenital tissues of female chimpanzees and on sexual behavior of the pairs. Elimination of the cyclic increase in anogenital swelling of the female is readily interpreted as resulting from the reduced ovarian secretion of estradiol, due to negative

Table 2. Social dominance and affiliation in FATs during the natural cycle and the mean rate of copulation per cycle in FATs and RATs combined, during natural cycles with a placebo (PL) and cycles with an OC

	Female dominance	Pair affiliation	Copulation	
			PL	OC
$Flora_{50}$ × Phineas	high	low	2	0
$Cheri_{400}$ × Frans	high	low	0	0
$Jacqueline_{400}$ × Rogger	moderate	moderate	2	0
$Anna_{50}$ × Harvey	moderate	high	19	0
Gay_{100} × Rogger	low	low	11	0
$Wenka_{50}$ × Frans	low	low	51	65
Li'l One_{100} × Phineas	low	moderate	45	22
$Sonia_{100}$ × Harvey	low	high	23	7
$Barbi_{400}$ × Harvey	low	high	25	5

A subscript after the female's name indicates the daily dosage of EE_2 (μg), combined with 0.5 mg norethindrone, the female received in the OC cycle.

Female dominance was based on the rate of female-to-male agonism, pair affiliation on the rate of allogrooming and copulation on the mean cycle rate for FATs and RATs combined.

feedback inhibition of pituitary gonadotropin secretion by the OC, combined with a local action of the progestin, norethindrone, on the anogenital tissues themselves. This interpretation is consistent with previous research on anogenital swelling in which swelling was induced in the ovariectomized female chimpanzee by treatment with an exogenous estrogen and detumescence was induced by a progestin [23]. The progressive increase in anogenital swelling with increasing dosages of EE_2, despite the progressive reductions in endogenous concentrations of estradiol, is consistent with the proposal that EE_2 has a greater biological action on the anogenital tissues than the natural estrogen.

The overall adverse effect of the OC on sexual behavior is likely an indirect effect, in part, of the reduced anogenital swelling. Since the female's anogenital swelling is a sexual attractant to the male, the reduced swelling during the OC cycle undoubtedly contributed to the lower rate of male sexual initiative under those conditions. The reduced proceptivity of the females during the OC cycle more likely resulted from reduced estrogenic stimulation at a central locus, though clearly, the behavior of any pair of chimpanzees is the result of a complex interaction between the female's hormonal condition and the behavior of the male partner. The attractiveness of the female is influenced by her behavior as well as her anogenital swelling, and the proceptivity of the female is influenced by the male's behavior toward her as well as her own hormonal condition [20].

Perhaps, the most interesting aspect of the research in terms of the possibilities for extrapolation to the human is the suggestion that the social relationship within the pair determined to a major extent both the rate of sexual activity during the natural cycle and the effect of the OC on sexual behavior. Although affiliation may have had some influence on sexual relations of the chimpanzees, its influence was most apparent by its absence in a single pair. Much more significant, apparently, was the dominance relationship between individual females and males, as reflected by the rate of female-to-male agonism. Those pairs in which the female's dominance was moderate or high had relatively low rates of copulation in the natural cycle and failed to copulate at all when the female received the OC. The critical factor in determining the nature of the sexual relationship was not the absolute level of agonism within the pair, though that likely played a role, but the relationship of the female's agonism to that of the male, here interpreted as an index of the female's agonistic dominance. Expressed somewhat differently, rates of sexual behavior in the natural cycle were relatively high and influenced less adversely by the OC in those pairs in which the male's dominance was not challenged by the female.

The implications of these results obtained from the chimpanzee for the human are twofold. The data on the female's anogenital tissues suggest that one relevant effect of the combined OC was to reduce stimulation of estrogen-sensitive peripheral target tissues, probably by the combined effect of reduced estrogen stimulation per se and an antagonistic action by the progestin. Although this type of effect is more pronounced in the female chimpanzee than it is in the woman, it likely has relevance to the woman by the following reasoning. The genital tissues that undergo swelling in response to estrogen stimulation in the chimpanzee are homologous to the labia minora of women [24, 25]. Adverse effects of reduced estrogen stimulation on the woman's genital tissues are clearly apparent when there is an extreme reduction in estrogen, such as following surgical or natural menopause [26, 27]. Since some women who use OCs report genital symptoms which are consistent with reduced estrogen stimulation [5], it is likely that the mechanism for those effects is similar to that described for the chimpanzee.

The effect of the OC on sexual behavior of the chimpanzees is interesting in several respects related to the chimpanzee as a model for the human. Although copulation was reduced to some extent in all the pairs when the female received the OC, the effect was profoundly influenced by the social relationship within the pair. Since social factors play a significant role in the sexual relations of humans, there is some reason to propose that data on the role of social factors on sexual behavior in the chimpanzees may have some relevance to the human. Similarly, although hormonal influences are less significant to the sexual behavior of humans than they are to chimpanzees, hormonal influences in humans cannot be discounted altogether. Given the great individual differences that exist among humans in all aspects of behavior, it is undoubtedly the case that hormones play a greater role in the sexual lives of some humans than they do in others.

Extrapolation of behavioral effects of OCs from the chimpanzee to the human requires consideration of both the relationship between the species with respect to hormone-behavior interactions generally and the nature of specific interactions in the research on OCs. As suggested above, with respect to hormonal influences on sexual behavior generally, they are less significant in humans than chimpanzees but more significant in some humans than in others. Results on OCs in the chimpanzee, therefore, at best represent a clue as to *possible effects* that would be less pronounced in humans overall but more pronounced, perhaps, in some humans than in others. Recognizing these limitations and qualifications on extrapolation to the human, the results on the chimpanzee suggest that OCs may have an adverse impact on human sexual behavior, but ultimately, the extent of the effect is likely to be minimal in individuals with an otherwise compatible social and sexual relationship. Possible qualifications to the predominant influence by social factors in determining the effect of OCs on human sexual behavior relate to individuals with an unusually high sensitivity to altered estrogenic stimulation in the woman, the clearest physiological effect of an OC. Such high sensitivity apparently occurs in some percentage of women, either with respect to the action of estrogen on the genitals [5] or at some central locus related more directly to behavior [4].

Not generally considered in the context of OC use in humans, but suggested by the results on the chimpanzees, is the further possibility that OCs could have an adverse effect on the sexual behavior of men, e.g. men who are highly sensitive to genital cues from their women partners. The reduced sexual initiative of the male chimpanzees was likely due, in part, to the reduced anogenital swelling of the females, i.e. due to an altered visual cue to the male regarding the female's state of sexual arousal. Since the genitals of the woman potentially provide a tactile cue to the man regarding the woman's state of sexual arousal, an alteration of that tactile cue by the OC could influence the man's sexual initiative in a manner analogous to that by which the altered visual cue of the female chimpanzee influenced the sexual initiative of the male chimpanzee.

The research on the chimpanzee suggests four mechanisms by which OCs may adversely influence human sexual behavior. The first three represent mechanisms with relatively low probability of occurrence in and of themselves. The fourth mechanism, related to the social and sexual relationship of the couple, represents a more likely possibility and one which is subject to experimental verification. High sensitivity to altered estrogen stimulation in the woman could alter female sexual motivation (1) directly at a central nervous system locus or (2) indirectly in response to genital discomfort during sexual intercourse [5]. (3) High sensitivity in the man to altered genital cues in the woman could adversely affect male sexual motivation. (4) Some combination of the conditions in [1–3] above, in couples that are not otherwise sensitive to such stimuli, could exacerbate in an additive way an already problematic sexual relationship. The interaction between the social relationship and hormonal effects of OCs, while seemingly intuitive post hoc, was clearly suggested by the research on the chimpanzees. The results, thereby, focus attention on biologically relevant variables involved in the use of OCs and further support the use of the chimpanzee as a model for the human in this general research area.

Conclusions

(1) Two aspects of sexuality in the chimpanzee were adversely affected by a combined OC: (a) female genital tissues and (b) sexual behavior of oppositely sexed pairs.

(2) The effect on the female chimpanzee's genitals appears to be comparable to genital effects reported for some women who use OCs. This research suggests that genital effects result from reduced levels of endogenous estradiol and the presence of exogenous progestin.

(3) Reduced copulation during OC administration was related to reductions in male sexual initiative and female proceptive behavior. Both reductions may be mediated indirectly by reduced estrogenic stimulation of the female's genitals, or reduction in female proceptive behavior could result from reduced estrogenic stimulation of a central neural mechanism.

(4) The social relationship of the pair was as important as, or more important than, the hormonal effects of the OC in determining the effects on sexual behavior.

(5) Given the less significant role played by hormones in the regulation of human sexual behavior, it is likely that social factors play an even greater role in the sexual responsiveness of humans to the woman's use of an OC.

(6) The value of the chimpanzee as a model for the human in this area of research is reflected in (1) relevant similarities and advantageous differences between chimpanzees and humans with respect to reproductive physiology and behavior, (2) clarification of a mechanism by which OCs may exert adverse effects on female genital tissues, (3) identification of a mechanism by which OCs may adversely influence male sexual behavior and (4) identification of variables within a social relationship that may contribute to sexual problems of couples in which the woman uses an OC.

Acknowledgments

This work was supported by grants HD-19060 from the National Institute of Child Health and Human Development, National Institutes of Health and BNS 91-09441 from the National Science Foundation to the author and grant RR-00165 from the National Center for Research Resources, National Institutes of Health, to the Yerkes Regional Primate Research Center of Emory University. The Yerkes Center is fully accredited by the American Association for Accreditation of Laboratory Animal Care.

References

1 James WH: Coital rates and the pill. Nature 1971;234:555–556
2 Dennerstein L, Burrows G: Oral contraception and sexuality. Med J Aust 1976;1:796–798.
3 Bancroft J, Sartorius N: The effects of oral contraceptives on well-being and sexuality: A review. Oxford Rev Reprod Biol 1990;12:57–92.
4 Hatcher RA, Guest FJ, Stewart FH, Stewart GK, Trussell J, Frank E: Contraceptive Technology 1984–1985. New York, Irvington Publishing, 1984.
5 Dickey RP: Managing Contraceptive Pill Patients. Durrant, Creative Informatics, 1983.
6 Bancroft J: The relationship between hormones and sexual behaviour in humans; in Hutchinson JB (ed): Biological Determinants of Sexual Behaviour. Chichester, Wiley & Sons, 1978, pp 493–519.
7 James WH: The distribution of coitus within the human intermenstruum. Science 1971;3:159–171.
8 Yerkes RM, Edler JH: Oestrus, receptivity and mating in the chimpanzee. Comp Psychol Monogr 1936;12:1–39.
9 Yerkes RM: Sexual behavior in the chimpanzee. Hum Biol 1939;11:78–111.
10 Young WC, Orbison WD: Changes in selected features of behavior in pairs of oppositely sexed chimpanzees during the sexual cycle and after ovariectomy. J Comp Psychol 1944;37:107–143.
11 Graham CE: Menstrual cycle of the great apes; in Graham CE (ed): Reproductive Biology of the Great Apes. Comparative and Biomedical Perspectives. New York Academic Press, 1981, pp 1–43.
12 Reyes FI, Winter JSD, Faiman C, Hobson WC: Serial serum levels of gonadotrophins, prolactin, and sex steroids in the non-pregnant and pregnant chimpanzee. Endocrinology 1975;96:1447–1455.
13 Nadler RD, Graham CE, Gosselin RE, Collins DC: Serum levels of gonadotropin and gonadal steroids, including testosterone, during the menstrual cycle of the chimpanzee *Pan troglodytes*. Am J Primatol 1985;9:273–284.
14 Dahl JF, Nadler RD, Collins DC: Monitoring the ovarian cycle of *Pan troglodytes* and *P. paniscus*: A comparative approach. Am J Primatol 1991;24:195–209.
15 Nadler RD, Dahl JF, Collins DC, Gould KG: Hormone levels and anogenital swelling of female chimpanzees as a function of estrogen dosage in a combined oral contraceptive. Proc Soc Exp Biol 1992;201:73–79.
16 Nadler RD, Dahl JF, Gould KG, Collins DC: Effects of an oral contraceptive on sexual behavior of chimpanzees *(Pan troglodytes)*. Arch Sex Behav, in press.
17 Briggs M, Briggs M: Plasma hormone concentrations in women receiving steroid contraceptives. J Obstet Gynaecol Br Commonw 1972;79:946–950.
18 Goldzieher JW, de la Pena A, Chenault CB, Cervantes A: Comparative studies of the ethynyl estrogens used in oral contraceptives. Am J Obstet Gynecol 1975;122:625–636.
19 Goldzieher JW, Chenault CB, de la Pena A, Dozier TS, Kraemer DC: Comparative studies of the ethynyl estrogens used in oral contraceptives: Effects with and without progestational agents on plasma androstenedione, testosterone, and testosterone binding in humans, baboons, and beagles. Fertil Steril 1978;29:388–396.
20 Beach FA: Sexual attractivity, proceptivity, and receptivity in female mammals. Horm Behav 1976;7:105–138.

21 Dixson AF: The hormonal control of sexual behaviour in primates. Oxford Rev Reprod Biol 1983; 5:131–219.
22 Nadler RD, Herndon JG, Wallis J: Adult sexual behavior: Hormones and reproduction; in Mitchell G, Erwin J (eds): Comparative Primate Biology: Behavior, Conservation, and Ecology. New York, Liss, 1986, vol 2A, pp 363–407.
23 Graham CE, Collins DC, Robinson H, Preedy JRK: Urinary levels of estrogens and pregnanediol and plasma levels of progesterone during the menstrual cycle of the chimpanzee: Relationship to the sexual swelling. Endocrinology 1972;91:13–24.
24 Hill WCO: The external genitalia of the female chimpanzee. Proc Zool Soc Lond 1951;121:133–145.
25 Wislocki GB: On the female reproductive tract of the gorilla, with a comparison to that of other primates. Contrib Embryol 1932;23:163–204.
26 McCoy N, Davidson JM: A longitudinal study of the effects of menopause on sexuality. Maturitas 1985;7:203–210.
27 Sherwin BB: Changes in sexual behavior as a function of plasma sex steroid levels in post-menopausal women. Maturitas 1985;7:225–233.

Ronald D. Nadler, PhD, Yerkes Regional Primate Research Center, Emory University, Atlanta, GA 30322 (USA)

Eder G, Kaiser E, King FA (eds): The Role of the Chimpanzee in Research.
Symp, Vienna 1992. Basel, Karger, 1994, pp 68–78

The Psychophysiology of Chimpanzee Perception

Gary G. Berntson[a], *Sarah T. Boysen*[a], *Michael W. Torello*[b]

[a] Department of Psychology, Ohio State University, Columbus, Ohio, USA;
[b] Yerkes Regional Primate Research Center, Emory University, Atlanta, Ga., USA

Questions as to the nature and origins of human intelligence and cognition have long occupied the attention of both philosophers and scientists. Although Cartesian dualism has given way to Darwinian concepts of the continuity of the human and animal mind, fundamental questions remain concerning the cognitive capacities of animals and the evolution of human intelligence. Differences between humans and animals in perceptual processes, behavioral repertoires and communicative skills severely limit empirical approaches to these questions and become especially restrictive for studies of infants. In view of these considerations, we have capitalized on inherent psychophysiological relationships between behavioral states and physiological processes, to gain insights into the ontogeny and phylogeny of cognition. Because of their unique evolutionary status, chimpanzees represent a pivotal species for the study of the evolution of mind. In the present chapter, we overview two psychophysiological approaches to the study of perception and cognition in the chimpanzee.

As is well documented, close functional links exist between neurobehavioral mechanisms and the autonomic nervous system. Autonomic adjustments constitute an integral aspect of adaptive response and are universally manifest in behavioral contexts. These psychophysiological responses can be highly informative as to the associated cognitive or behavioral states, when employed in creatively designed experimental paradigms [1]. An additional psychophysiological approach capitalizes on the electrical signals generated by neural processing of information. Event-related brain potentials (ERPs) reflect not only early stages of sensory transmission but subsequent perceptual and cognitive reactions as well. Importantly, heart rate and ERP measures are broadly applicable across age and species and are not dependent on the behavioral repertoire or communicative skills of the organism. Because they represent the aggregate activities of neural systems, psychophysiological measures are often more closely linked to behavioral and cognitive processes than are more molecular or cellular measures.

The psychophysiological approach is illustrated by our earlier studies of photograph recognition in the chimpanzee. Although it is well established that animals can learn to discriminate photographs of faces, it is not at all clear if animals in fact recognize the individuals depicted. That is, whether their perceptual reactions entail a cognizance of the individual characteristics and defining features of the depicted individual. To explore this issue, we examined heart rate responses of a chimpanzee to facial photographs of humans and conspecifics [2, 3]. The subject was a 3.5-year-old female chimpanzee (Sheba), who had formed strong social bonds with her two primary human caregivers and had established stable relationships with conspecifics in the course of involvement in the OSU Primate Cognition Project. To examine the impact of these established social relationships on Sheba's perceptual response, facial photographs of humans and chimpanzees were repeatedly presented, while concurrent heart rate responses were obtained. We initially examined Sheba's responses to slide presentations of facial photographs of her primary human caregivers, unfamiliar individuals and familiar individuals with whom Sheba had not interacted [2]. In a subsequent study, we also tested facial photographs of conspecifics, including a familiar playmate, an unfamiliar animal and a familiar animal who had been consistently aggressive to Sheba [3]. Importantly, Sheba had no previous experience with these photographs, no explicit training was given in either facial or photograph recognition, and no task demands were imposed. Sheba merely observed sequences of slide stimuli. Only modest heart rate responses were observed to slide photographs of individuals who had no established social relationship with the subject. Slide presentations of human caretakers, however, evoked striking heart rate *decelerations*, characteristic of the orienting response to attentionally significant stimuli [1]. In marked contrast, photographs of the aggressive chimpanzee yielded significant heart rate *accelerations*, reminiscent of defensive responses [1].

These differential cardiac responses document Sheba's ability to recognize significant individuals from photographic representations, in the absence of explicit training, task demands or prior experience with the photographs. Because we equated physical parameters of the slide stimuli, such as intensity and color, differences in the pattern of cardiac response to the facial photographs could be attributed only to the social significance of the individuals depicted. These results illustrate the sensitivity of psychophysiological measures to patterns of perceptual response that would otherwise have gone unnoticed, since no behavioral responses to the stimuli were observed.

The Ontogeny of Cardiac Reactions to Vocal Signals

The differential reactions to faces, as discussed above, were based on social learning and the established social relationships of the subject. In other cases, however, adaptive responses to certain classes of stimuli may be constitutionally endowed. Infant cries, for example, may inherently evoke affective responses in both humans and nonhuman primates [4, 5]. In view of the importance of vocal communication in primates, it is likely that both humans and chimpanzees are endowed with specialized perceptual mechanisms, tuned to specific aspects of vocal signals. Moreover, these mechanisms may be inherently linked to behavioral mechanisms, permitting an early ontogenetic emergence of adaptive response. To explore this issue, we employed heart rate measures to examine the ontogeny of vocal perception in apes [6].

Briefly, epochs of chimpanzee vocalizations were presented to infant and neonatal chimpanzees and orangutans, while concurrent heart rate measures were obtained. Subjects were 6 chimpanzees (48 h to 2.9 years of age) and 3 orangutans (2.4–5.0 months old) included for comparison. With the exception of the oldest chimpanzee, who was a participant in the Primate Cognition Project at OSU, all animals were housed under comparable conditions in the nursery of the Yerkes Regional Primate Research Center (Emory University, Atlanta). For a variety of reasons, all animals were separated from their mothers at or shortly after birth. The animals were tested for cardiac reactivity to experimental stimuli, which consisted of 1-second epochs of white noise, chimpanzee threat barks, distress screams and low-level alarm calls. The intensity of all stimuli was equated to 76 dB (SPL). All animals were tested while in a quiet wakeful state. Six trials with each stimulus were given in block-randomized order and heart rate measures were taken before, during and after each stimulus presentation.

Infant orangutans displayed deceleratory heart rate responses to all acoustic stimuli. Cardiac deceleration is a characteristic feature of the orienting response and likely reflected the general attentional significance of the stimuli. As illustrated in figure 1, infant chimpanzees also evidenced predominantly deceleratory heart rate responses to most stimuli. In response to conspecific threat barks, however, chimpanzees displayed an opposite, cardioacceleratory response. This species-specific cardiac acceleration to threat barks was significantly different from the response to any other stimulus and is highly reminiscent of the defensive reaction to pain or fear-arousing stimuli [1, 7]. This cardioacceleratory response was seen consistently across subjects and was apparent within the early neonatal period, even for animals that had been isolated in the nursery from birth. Indeed, the youngest animal (48 h old) displayed among the largest acceleratory responses to the threat vocalization.

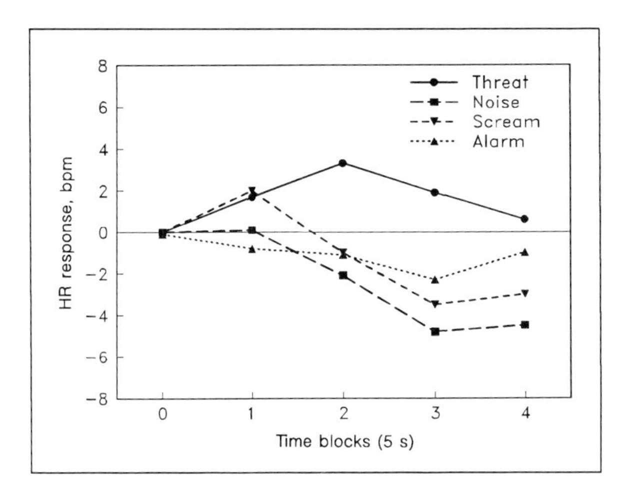

Fig. 1. Heart rate (HR) responses of chimpanzees (n = 6) to 1-second epochs of conspecific vocalizations and a white noise control stimulus. Experimental stimuli were presented at time 0 and data points illustrate the mean heart rate response over subsequent 5-second time blocks.

Although an acceleratory response to threat barks was observed in each chimpanzee, none of the orangutans evidenced this pattern of response. These findings are consistent with the view that chimpanzees are constitutionally endowed with specialized perceptual mechanisms, tuned to specific features of conspecific vocalizations and inherently linked to patterns of adaptive response.

ERPs and Vocal Perception

Acoustic perception of nonhuman primates evidences features similar to those which underlie speech processing in humans. These include the categorical perception of continuously graded phonemes and the lateralized hemispheric processing of specific vocal signals [8–12]. In view of these considerations, a further understanding of vocal processing in nonhuman primates may enhance our understanding of the mechanisms of vocal communication in these species and offer insights into the evolution of language processes. One approach to the study of neural processing of sensory stimuli is through the application of ERPS. These have been shown to be sensitive to cognitive and attentional processes and the topographic features of the evoked waveform and its scalp distribution can offer insights into the nature of the underlying perceptual processing. Initial components of the evoked response (<50 ms) are associated with the early stages of sensory transmission and are often related to the specific physical characteristics of the stimuli. In many cases the specific neural generators of these early components have been extensively detailed [13]. In contrast, later components of the evoked waveform are not as rigidly fixed but covary with patterns of cognitive processing of the stimuli. An example of the latter is the P300, a positive-going potential with a typical latency between 300 and 500 ms. The P300 is distributed primarily in the central scalp region and its amplitude has been shown to be closely related to the informational significance of the stimulus. In addition to the classical P300, there now appears to be a family of late positive components (the late positive complex), with different latencies, waveforms, scalp distributions and patterns of sensitivity to cognitive events [14, 15].

We have previously demonstrated that visual ERPs of chimpanzees are highly similar to those of humans and evidence parallel developmental changes [16]. More recently, we employed ERP measures to explore the perceptual processing of vocal signals in a chimpanzee (Sheba). Sheba was lightly sedated with droperidol and ketamine for electrode application and testing but remained attentive and behaviorally responsive to environmental stimuli throughout. An array of 28 scalp and 4 artifact electrodes was applied by standard convention and time-locked electroencephalographic activity was recorded to discrete (300–400 ms) acoustic stimuli (2-second interstimulus intervals). Stimuli included pure tones, white noise and brief epochs of human speech and conspecific vocal stimuli (threat barks, screams and alarm calls as tested above). Each ERP run consisted of at least 20 artifact free recordings for a given stimulus. Stimuli were delivered under computer control and time-locked event-related brain activity was recorded and processed by a Neuroscience ERP system (2 ms sample, 16 ms prestimulus and 600 ms poststimulus times). Artifacts were automatically rejected and the remaining traces were averaged separately for each electrode site. In all cases, results were replicated in two or more runs to ensure reliability of the evoked response.

To extend findings from humans and to confirm the attentiveness of our subject, we first tested Sheba in a standard P300 'oddball' paradigm. In humans, a late positive (P300) component emerges in the ERP response to an infrequent stimulus (e.g. a high-frequency tone) embedded in a stream of frequent stimuli (e.g. low-frequency tones). The P300 has been shown to be related to the novelty and informational significance of the rare tone [17]. Although the P300 is substantially enhanced by imposing a response demand on the occurrence of the rare stimulus (e.g. counting, pressing a button), typical P300 responses can be seen even in the absence of an explicit response [18]. No task demand was imposed in the present study. The stimuli consisted of pure

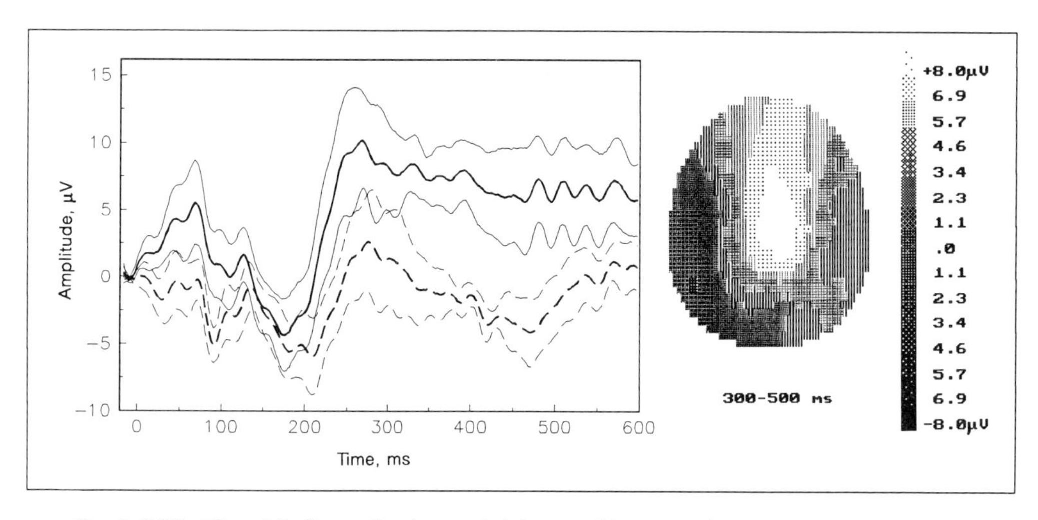

Fig. 2. ERPs (Cz, oddball paradigm) recorded from a chimpanzee in a standard P300 paradigm. The thick lines illustrate the average ERP waveforms (80 traces) to the rare (——) and frequent (---) stimuli. Thin lines depict the results from the two separate sessions (40 traces). Note the late positivity occurring to the rare stimulus, relative to the frequent stimulus. The topographic map (right) illustrates the scalp distribution of the difference waveform (ERP to the rare stimulus – the ERP to the frequent stimulus, from 300 to 500 ms). Top is anterior, bottom is posterior. Note the predominant central or fronto-central distribution of the later positivity.

tones of two frequencies (500 and 1,500 Hz) presented at 2-second interstimulus intervals, with the probability of one stimulus being 85% and the other 15%. Two sessions were given, with a total of 40 traces acquired for the rare and 40 for the frequent stimulus in each session.

Results are illustrated in figure 2, which depicts the ERPs to the rare and frequent stimuli for each session (thin lines), and the overall average (80 traces, heavy lines). Consistent with the literature on humans, the rare stimulus was associated with a late positive deflection that became most apparent about 350 ms after stimulus onset. Beyond this point, there was no overlap in the ERP waveforms to the rare and frequent stimuli. Further consistent with findings in humans, the late positivity was focussed around the central scalp (Cz). This is illustrated in the topographic map of figure 2, which shows the scalp distribution of the difference waveform (ERP to the rare stimulus – the ERP to the frequent stimulus). These results confirm the general similarity of ERPs in humans and chimpanzees and the sensitivity of these measures to stimulus probability.

Stimulus meaning or significance can enhance long latency potentials in humans (e.g. the P300 [17]), and this is of particular importance for studies of

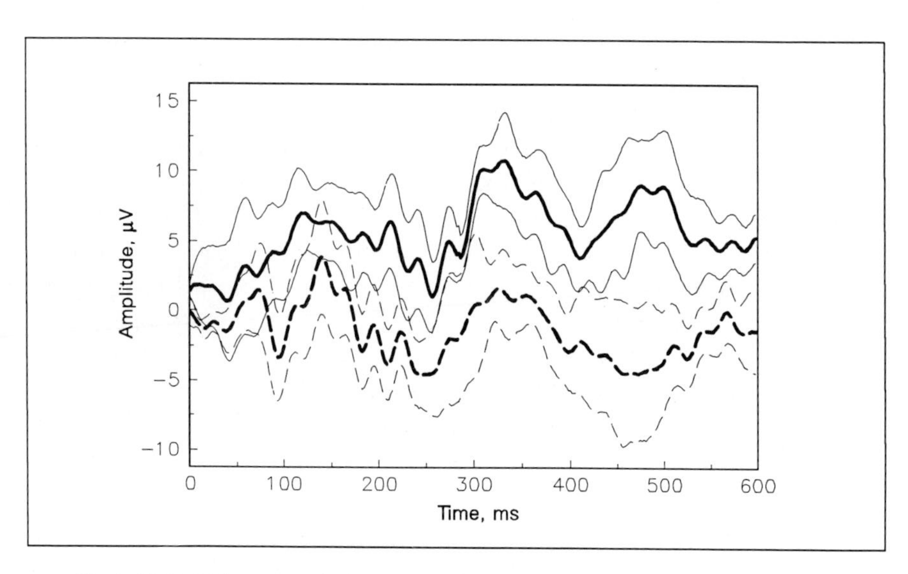

Fig. 3. ERPs (Cz) to a vocal stimulus (——) and a control tone (- - -). The vocal stimulus consisted of the animal's name (Sheba) and the control tone was a 500-Hz tone. Thick lines illustrate the average ERP waveforms (100 traces) to the two stimuli. Thin lines depict the standard errors of the waveforms across 5 separate 20-trial blocks. Note the late positivity occurring to the vocal stimulus, relative to the control tone.

vocal perception. To examine the sensitivity of the ERP to stimulus meaning in the chimpanzee, we recorded brain ERPs to two stimuli differing in adaptive significance to the animal. Sheba again served as the subject and the stimuli were a 400 ms (500 Hz) tone and a 400 ms epoch of human speech. The speech stimulus consisted of a single iteration of the animal's name (Sheba), spoken by her primary caretaker. The stimuli were digitized and replayed under computer control. Five blocks of 20 artifact-free traces were obtained for each stimulus. Only a single stimulus type was presented within a given block so that no biases could arise from differences in stimulus probability. The order of stimulus administration was counterbalanced across blocks. Other procedural details were as outlined above. Figure 3 illustrates the overall average ERPs for the two stimuli (heavy lines), together with the standard errors of the waveforms across the 5 replications for each stimulus (thin lines). The ERP to the vocal stimulus ('Sheba') displayed a late positive deflection, relative to the control stimulus. While there was considerable overlap in the ERP distributions for the two stimuli in the early poststimulus period, the traces diverged after about 300 ms. Beyond that point, there was no overlap between the two ERP waveforms or their standard errors. Topographic analysis again revealed a predominant central-scalp locus of this late positivity, with a maximum at Cz. This positivity was somewhat more frontally distributed like the P300-like component in the oddball paradigm.

The neural origins and functional significance of the late positive potential to the vocal stimulus remain to be determined. Although it shares some features with the P300-like response to the rare stimulus obtained above, the late positive potential was not obtained in a typical P300 paradigm. There were no differences in the relative probabilities of the stimuli. Moreover, the late positive potential appears to have a somewhat more frontal distribution than does the P300-like response to the rare stimulus. Regardless of the ultimate resolution of these issues, the late positive component of the ERP appears to reflect the adaptive significance of the vocal stimulus. In view of these findings, we further investigated ERPs to the conspecific vocalizations (threat, alarm, scream, white noise) as tested above with heart rate measures. These studies addressed several issues. First, results with conspecific vocalizations could clarify whether the late positive potential observed above was specific to human speech (or the specific stimulus word) or is a more general manifestation of the processing of vocal signals. Secondly, the inclusion of a broad-band white-noise stimulus permits an evaluation of the possibility that the late positive potential to the vocal stimulus, relative to the tone, was attributable simply to the wider spectral frequency composition of the former. Finally, these ERP studies may reveal subtle differences in brain processing of the conspecific vocal stimuli which yielded divergent patterns of heart rate response in the studies outlined above.

General procedures followed those described above. Briefly, 300-ms epochs of white noise as well as conspecific threat, alarm and scream vocalizations were presented in separate blocks. A single stimulus was presented during a given block, with at least 20 artifact-free traces acquired. Each stimulus was presented in two separate blocks, with the order of stimuli counterbalanced across blocks. The scream and alarm stimuli yielded virtually identical responses and were thus collapsed for illustration. Figure 4a shows the ERP waveforms to threat contrasted with the other vocalizations. The threat stimulus evoked a slow positivity that persisted for most of the 600-ms acquisition window. Consistent with the distinct cardiac responses to this stimulus, the late positivity to threat barks was appreciably greater than for other vocal stimuli. Indeed, the late positive component (beyond 300 ms) to threat was about twice as large as the response to the other vocal signals. Consistent with the results obtained above with human vocalizations, the topography of this late positive potential evidenced a predominant frontocentral distribution, with a maximum at Cz. The ERP to threat was also more positive in the early portion of the waveform (<300 ms), although this may have been due to differences in the rise times of the vocal signals, which can affect ERP components. Because the

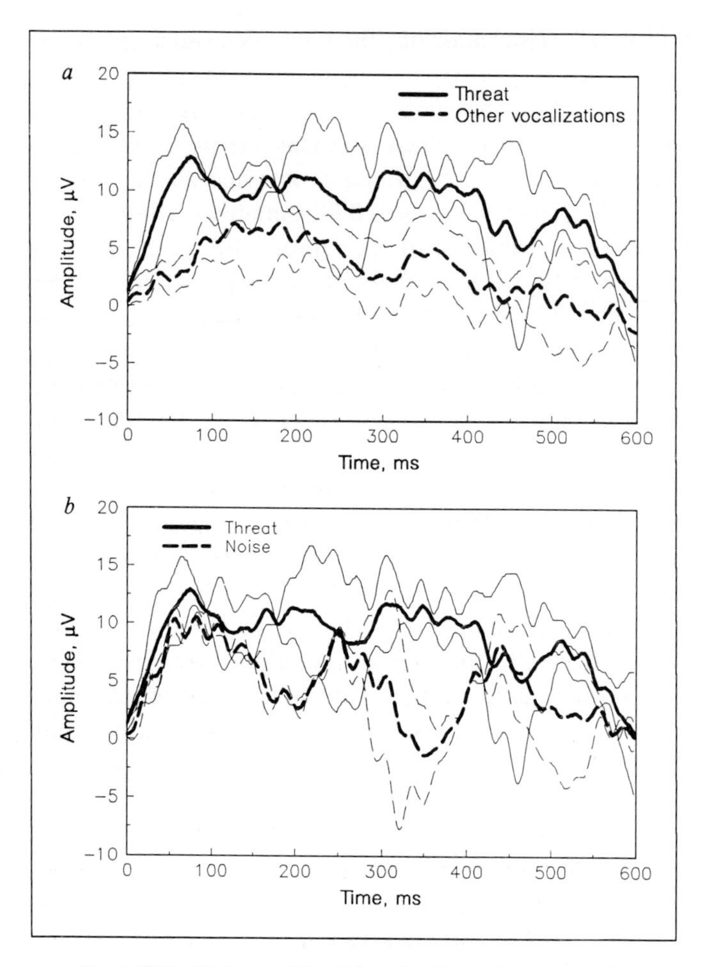

Fig. 4. ERPs (Cz) to a white-noise stimulus and to conspecific vocalizations. Responses to screams and alarm calls were virtually identical and are collapsed for illustration *(a)*. Thick lines depict the overall average ERP waveforms to the experimental stimuli (40 traces per stimulus). Thin lines depict the two separate 20-trial acquisitions (threat and noise, *b*) or the standard error of the waveform (other vocalizations, *a*). Note the enhanced late positivity occurring in response to the threat vocalization, relative to other stimuli.

rise time of the vocal envelope can be a critical acoustic feature of the vocal signal, we could not simply taper these onsets without potentially altering the communicative significance of the vocalizations. Because of these differences in rise time, however, we also employed a rapid rise time for the white-noise control stimulus. As apparent in figure 4b, the noise stimulus also yielded an early positive-going deflection that closely approximated that to the threat stimulus. Consistent with the findings for other vocalizations, the later components of the ERP to threat were generally more positive than for noise, especially in the 300 to 400 ms range.

Conclusion

The present results reveal that psychophysiological responses can be highly sensitive to cognitive variables, even in paradigms which do not require training or behavioral responding. Psychophysiological measures are broadly applicable across ontogeny and through phylogeny and offer important approaches to the study of perception and cognition that do not depend on the behavioral involvement of the subject. Results discussed above indicate that the adaptive significance of a stimulus is an important determinant of psychophysiological response. While some patterns of psychophysiological response may be attributed to social learning (e.g. responses to faces or names), others (e.g. responses to threat) emerge early in ontogeny and may be constitutionally endowed. Thus, both heart rate and ERP measures distinguished between threat barks and other conspecific vocal signals. This differentiation was species specific and became apparent within 48 h of birth. These findings are consistent with the existence of special perceptual mechanisms for vocal signals in chimpanzees, which may inherently bias social and environmental perception. In view of the probable phylogenetic continuity of such mechanisms and their likely appearance in humans, it would be of fundamental importance to clarify the inherent biases that color our perception of the world. The strategic application of psychophysiological measures holds considerable promise for illuminating this issue and for explorations of the evolution of intelligence and the relationships between human and animal minds.

Acknowledgement

This work was supported in part by an NSF grant (BNS-9022355) to S.T.B., an NIMH grant (MH-45460) to G.B.B. and by NIH Division of Research Resources Grant RR-00165 to the Yerkes Regional Primate Research Center. The Yerkes Center is fully accredited by the American Association for Laboratory Animal Care.

References

1 Berntson GG, Boysen ST: Cardiac indices of cognition in infants, children and chimpanzees; in Rovee-Collier C, Lipsitt L (eds): Advances in Infancy Research. New York, Ablex, 1990, vol 6.

2 Boysen ST, Berntson GG: Cardiac correlates of individual recognition in the chimpanzee *(Pan troglodytes).* Comp Psychol 1986;100:321–324.
3 Boysen ST, Berntson GG: Conspecific recognition in the chimpanzee: Cardiac indices of significant others. J Comp Psychol 1989;103:215–220.
4 Lester B, Newman J (eds): Biological and Social Aspects of Infant Crying. New York, Plenum Press, 1993.
5 Newman JD: The infant cry of primates: An evolutionary perspective; in Lester BM, Bokydis CPZ (eds): Infant Crying: Theoretical and Research Perspectives. New York, Plenum Press, 1985, pp 307–323.
6 Berntson GG, Boysen ST: Specificity of the cardiac response to conspecific vocalizations in chimpanzees *(Pan troglodytes).* Behav Neurosci 1989;103:235–245.
7 Berntson GG, Boysen ST, Bauer HR, Torello MS: Conspecific screams and laughter: Cardiac and behavioral responses of infant chimpanzees. Dev Psychobiol 1990;22:771–787.
8 Kuhl PK: The special-mechanisms debate in speech research: Categorization tests on animals and infants; in Harnad S (ed): Categorical Perception for Speech. London, Cambridge University Press, 1987, pp 355–386.
9 Morse PA, Molfese D, Laughlin NK, Linnville S, Wetzel F: Categorical perception for voicing contrasts in normal and lead-treated rhesus monkeys: Electrophysiological indices. Brain Lang 1987;30:63–80.
10 Petersen MR: The perception of species-specific vocalizations by primates: A conceptual framework; in Snowdon CT, Brown CH, Peterson MR (eds): Primate Communication. London, Cambridge University Press, 1982, pp 171–211.
11 Petersen MR, Zoloth ST, Beecher MD, Green S, Marler PR, Moody DB, Stebbins WC: Neural lateralization of vocalizations by Japanese macaques: Communicative significance is more important than acoustic structure. Behav Neurosci 1984;98:779–790.
12 Pohl P: Central auditory processing. V. Ear advantages for acoustic stimuli in baboons. Brain Lang 1983;20:44–53.
13 Regan D: Human Brain Electrophysiology. New York, Elsevier, 1989.
14 Ruchkin DS, Sutton S, Mahaffey D: Functional differences between members of the P300 complex: P3E and P3b. Psychophysiology 1987;24:87–103.
15 Sutton S, Ruchkin DS: The late positive complex. Ann NY Acad Sci 1984;425:1–23.
16 Boysen ST, Berntson GG: Visual evoked potentials in the great apes. Electroencephalogr Clin Neurophysiol 1985;62:150–153.
17 Johnson R: The amplitude of the P300 component of the event-related potential: Review and synthesis; in Ackles PK, Jennings JR, Coles MGH (eds): Advances in Psychophysiology. Greenwich, Jai Press, 1988, vol III, pp 69–137.
18 Polich J: Comparison of P300 from a passive tone sequence paradigm and inactive discrimination task. Psychophysiology 1987;24:87–103.

Gary G. Berntson, Department of Psychology, Ohio State University,
48 Townshend Hall, 1885 Neil Avenue, Columbus, OH 43210 (USA)

Eder G, Kaiser E, King FA (eds): The Role of the Chimpanzee in Research. Symp, Vienna 1992. Basel, Karger, 1994, pp 79–86

The US Chimpanzee Breeding and Research Program

Thomas L. Wolfle[a], *Milton April*[b]

[a] Institute of Laboratory Animal Resources, National Research Council, National Academy of Sciences, Washington, D.C., USA;
[b] Primate Research Centers Program, National Center for Research Resources, National Institutes of Health, Bethesda, Md., USA

More than 6 years before human acquired immune deficiency syndrome (AIDS) was identified, US biomedical chimpanzee biologists began to be concerned about the future of US chimpanzee breeding colonies. Chimpanzee biologists and a committee of representatives from federal agencies that use nonhuman primates in research, the Interagency Primate Steering Committee (IPSC), began a series of meetings to evaluate the status of captive chimpanzees and their chances for survival as a self-propagating population. The US captive chimpanzee population was in danger of extinction and importation from the wild had not been a viable option after the Convention on International Trade in Endangered Species (CITES) had been signed in 1975. By 1984 it was clear that the survival of the US captive chimpanzee population was unlikely without dedicated breeding efforts. A recommendation was made to the IPSC and in turn to the National Institutes of Health (NIH), to establish a national chimpanzee breeding program. The NIH recognized the value of the chimpanzee in AIDS research and provided the support needed to launch the National Chimpanzee Breeding and Research Program. The purpose of this presentation is to summarize the convergence of these forces and their impact on AIDS research and the captive chimpanzee population.

In the 1960s, hundreds of thousands of nonhuman primates were imported into the US primarily for use in the development, testing and manufacture of polio vaccine. Throughout the 1970s, the availability of nonhuman primates for research decreased due to the destruction of natural primate habitats and embargoes by source countries. Also, many species of nonhuman primates were categorized as endangered and all were listed in the first two appendices of the CITES. The decreasing supply caused biomedical investigators to be concerned about the long-term availability of nonhuman primates for research.

In 1974, the Public Health Service established the IPSC '... to develop a uniform approach to assure both short- and long-term supplies of nonhuman primates for bioscientific purposes in the United States' [1, p. iii]. Within a year, an international group of chimpanzee biologists and ecologists were convened at the NIH to recommend strategies for ensuring adequate supplies of chimpanzees. Their recommendation to establish a captive breeding program reinforced the existing concern about availability and set into motion a series of other meetings among scientists, conservationists and others interested in assuring the survival of the chimpanzee. The resulting National Primate Plan estimated the annual research need of 180 chimpanzees and labeled the chimpanzee 'the irreplaceable model for the study of certain human health problems' [1, p. 62]. It recommended that research sponsors consider supporting a nationally coordinated breeding program and that countries in which chimpanzees were threatened to be killed as agricultural pests should rescue offending animals and establish breeding colonies within their native country.

Under the sponsorship of the IPSC, annual meetings were held with chimpanzee biologists, behaviorists, physiologists and colony managers to clarify the goals and needs of a breeding program. It became immediately apparent that little was known about the demographics of the US chimpanzee population and whether it could sustain long-term breeding without the introduction of wild-caught animals. Ulysses S. Seal and Nate Flesness of the International Species Information System (ISIS) contributed a knowledge of inbreeding coefficients and population dynamics from a national perspective and for the first time observations were made across all participating colonies. Production data, mortality and record systems were compared. Definitions of 'stillborn', 'neonate', 'fetal wastage', and 'live birth' were developed and standardized. Open discussion of these issues was difficult because comparisons among colonies sometimes led to criticisms of management. Besides teaching demographics, ISIS taught colony managers to work together for the greater good of chimpanzees. Cohesion developed and a plan began to take shape.

Figure 1, developed by ISIS in the mid-1980s, depicts an age pyramid of the approximately 1,000 chimpanzees in the US research colonies. Males are shown on the left, females on the right, ages on the vertical axis and the numbers of chimpanzees are on the horizontal axis. The stippled top of the pyramid represents the almost exclusively wild-caught breeding population. The hatched bottom of the pyramid represents captive-born infant and juvenile

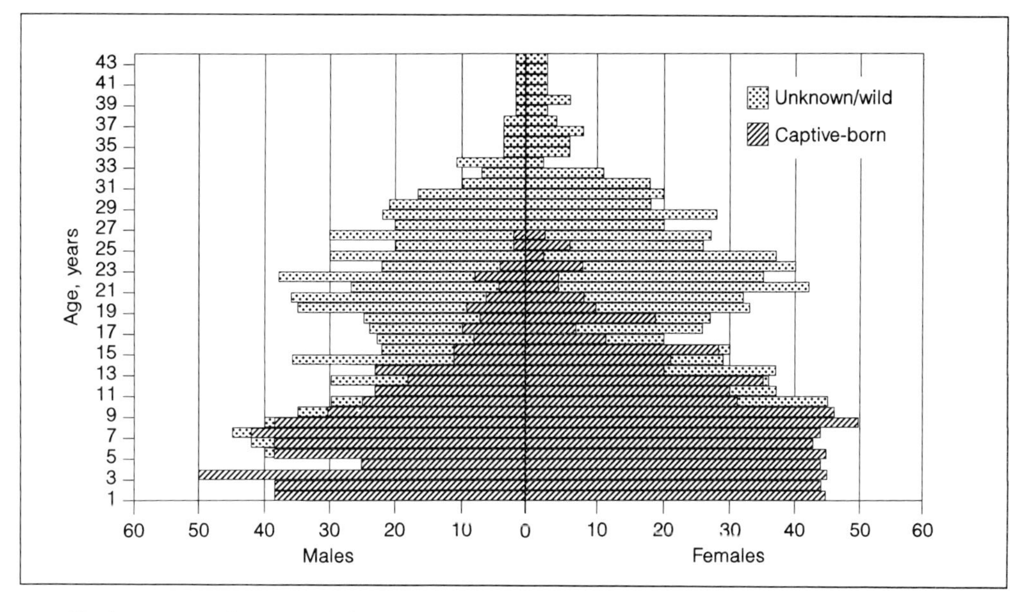

Fig 1. Age pyramid report. Chimpanzee studbook of the 1980s.

animals that are destined to become future breeders [Nate Flesness, pers. commun.].

As captive-born animals moved up the pyramid to replace their wild-born parents, the composition of the population was destined to become increasingly captive born. From what was already known about captive-born animals, it was anticipated that without major changes in the management of young animals, these animals would become increasingly unable to breed and rear young.

By the early 1980s, the wild-caught animals were nearing the end of their productive breeding life, and relatively few behaviorally healthy young animals were in the pipeline to become breeders. The breeding and parenting capacity of the young captive-born animals was highly suspect. Most had been separated from their mothers at birth and hand reared for use in hepatitis research without learning important social, reproductive and parenting skills. ISIS predicted that within 2 decades breeding efforts would begin to collapse without immediate action to ensure future breeders. Without the ability to import new animals, this meant that rather dramatic changes had to be made in the rearing methods for chimpanzees.

The practice of hand-rearing infant and juvenile animals was identified as a major contributor to the problem. However, changes in this practice necessitated major changes in research uses of chimpanzees, which up to that point had depended largely on the use of young animals. In addition to the problem of this reproductively inept cohort, hepatitis infections and mortality from nonexperimental causes would disqualify nearly 50% of the offspring from being included in dedicated breeding colonies.

From these data, ISIS projected that US nonzoo chimpanzee reproduction would begin a fatal decline in the 1990s, ending in zero production shortly after the turn of the century. This declining population curve corresponded with other captive populations that ISIS was tracking and established a sense of urgency.

The IPSC's ad hoc chimpanzee task force, as the group came to be known, accepted the challenge and in 1980 began a series of annual meetings at the University of Texas Cancer Center Science Park, the Yerkes Regional Primate Research Center in Georgia, the New Mexico Primate Research Institute, the Primate Foundation of Arizona and a number of other sites.

Guided by an initial report of the task force in 1980 [2], three subgroups – Behavior, Demographics and Reproductive Physiology – were established to develop guidelines and priorities for management and research that would be needed to establish a national breeding program. Research was needed to better understand how long offspring must spend with their parents to become good breeders and parents themselves. Other problem areas also needed to be addressed, such as increasing male sexual performance and fertility, effectively using artificial insemination and semen banking, reducing infant mortality, providing environmental enrichment and optimizing breeding and management systems.

To improve their ability to assess demographic data and make specific recommendations on the size and structure of a breeding population, the work of the task force was guided by the recommendations of the subgroups. These recommendations included using biological markers for demographic analysis, developing a handbook on chimpanzee colony management and promoting multiple use and noninvasive nondestructive experimentation. The task force also recommended the identification of breeder populations either known or believed to be hepatitis virus free. With the assistance of ISIS, the task force also recommended that a unique identification number be assigned to each chimpanzee.

The Behavior subgroup drafted a handbook on chimpanzee management at the Primate Foundation of Arizona and presented it for review by the task force in 1982. At this time, the chimpanzees with breeding potential began to be identified. Agreement on what constituted 'breeding potential' required extensive discussion about management and research practices, as well as research uses that would most likely lead to behaviorally and physiologically normal animals free from communicable diseases or infectious agents.

In 1983, ISIS was asked to survey the number of chimpanzees in the USA and break them down into four groups that ranged from those not at risk (no possible exposure to hepatitis agents) to those with known risk of infection through experimental exposure or contact with infected animals. From this survey, ISIS determined that there were approximately 1,235 chimpanzees in the USA, 90% of which were supported by biomedical research. Of these, 200 animals were known to have been used in non-A non-B hepatitis research. This was a troublesome problem at the time because there were no suitable diagnostic tests to identify secondarily infected animals and assess the risk of vertical or horizontal transmission. The devastation of nonsocial rearing became clearly evident when ISIS determined that of the 208 proven breeders in the US research population (145 females and 63 males), only 17 females and 4 males were captive born. No captive-born male was a grandfather! The lesson kept reappearing: The US captive breeding population was indeed in jeopardy.

ISIS concluded that a 100-year self-sustaining population should begin with 350 animals and should include a predetermined number of animals in each age cohort. Amazingly, when each of the existing 1,200 chimpanzees in the USA were matched against the agreed-upon criteria to be used in selecting breeders (e.g. healthy; physiologically sound; virus-free; socially reared; not currently, or previously, on a research protocol that would preclude breeding; proven breeder), approximately 350 animals were identified. Using fecundity and mortality data from an existing major US colony, ISIS projected that a colony beginning with 350 breeders could increase its population by 10% annually and be self-sustaining at half this rate of production. This meant that 50% of the newborns could remain in their natal groups as future breeders and 50% could be removed from their mother at birth (thereby shortening her intergestational period) and used in research at an appropriate age.

This chronology does not imply that sound practices of husbandry and research awaited the recommendations of the task force or endorsement by the IPSC. On the contrary, for a number of years most colonies had been employing improved husbandry and breeding strategies in which nursery rearing began to focus more on the socialization needs of the infant.

It became clear that for the US chimpanzee population to become self-sustaining over a long period, research programs using chimpanzees would be the only potential sources of financial support [3, 4]. Additional help was sought from zoological institutions for advice and contributions to the gene pool. Fortunately, ISIS monitors both the biomedical and zoological populations.

ISIS and task force participants identified problems to be addressed prior to initiation of a dedicated breeding program and gathered information about the established population. Hundreds of hours were devoted to soliciting applications, developing and building a program, justifying budget requests and establishing ground rules for a cooperative breeding program that encompassed existing colonies.

The most difficult issue was and continues to be, how to separate breeders and their progeny from animals returning from research in which they were exposed to hepatitis viruses and other diseases. Inadequate space posed a critical problem, new facilities appeared unlikely and euthanasia was unacceptable. In addition, birth control was considered incompatible with the objective of producing an increased number of socially reared future breeders, while assuring that medical research needs for chimpanzees were met – often impossible to anticipate several years in advance.

Convinced of the need for a national breeding program, yet stymied by lack of funding, the breeding plan languished for more than a year until the task force presented its recommendations to the IPSC in 1984. The timing of the program could not have been better. With the increased importance of AIDS came a recognition by the NIH that chimpanzees would become increasingly important for this research and AIDS funds became available for the program. Following further review, the NIH approved the plan to make competitive awards for the National Chimpanzee Breeding and Research Program [5]. About $ 4.5 million was allocated in 1986 to begin the program. Five awards for breeding and five awards for research were made.

Postscript on a Negative Side

Overcrowding and long-term support of aging chimpanzees infected with agents such as hepatitis and HIV and of no further value for research present a serious practical and ethical dilemma. While euthanasia is an unacceptable management tool by which to create space, funding to meet increased needs for housing seems unlikely. The only immediate option seems to be to curtail breeding – a process currently being initiated in several colonies.

To date, the most innovative idea for addressing this problem in the long-term is to establish endowments for animals that are used in research projects that may affect their future usefulness. Investigators who use animals are asked to pay a negotiated fee, which, when invested with contributions from other investigators, creates an endowment for long-term care of the group.

Postscript on a Positive Side

The NIH Chimpanzee Breeding and Research Program's chimpanzee population has grown by 42% since 1986. There are currently a total of 498 animals, consisting of 278 adults, 41 adolescents and 180 juveniles and infants. To date, there have been 288 live births! Some new facilities have been built and the practice of leaving offspring with their family group has become basic policy. Research in chimpanzee behavior, genetics, breeding and reproduction continues to produce important new information. The goal of fulfilling research needs, while also assuring the existence of future generations of US breeder chimpanzees, is being achieved. It is important to note that no feral chimpanzees have been imported into the USA for research since 1977. The success of this program has resulted in the firm commitment that no additional chimpanzees will be imported from the wild for research.

The lesson learned once again is that while the need for animal homologs of new human diseases can be generated almost overnight, the production of these homologs may require years and in this case as long as a decade. During the work of the task force, there was no funding in hand for support of the program, yet those closest to the problem and to chimpanzees knew that a solution had to be found. If the epidemic of AIDS had not occurred, this program might not have been possible. When AIDS is cured and the immediate need for chimpanzees becomes less acute, we need to continue to ensure the survival of the world's captive chimpanzee population. Furthermore, we need to extend this concern and support to other animal species that may provide unique and valuable animal models in their own right but lack the enormous public empathy there is for chimpanzees.

Dedicated breeding programs like this one and others for nonhuman primates established in countries of origin and supported because of their value to biomedical and health research, offer a real promise for assuring that there will be a future for chimpanzees and other endangered populations.

In summary, the managers of US chimpanzee breeding colonies, research scientists and the US Public Health Service have successfully worked together to ensure the survival of *Pan troglodytes* and have provided a vitally needed model for the study of AIDS and other important diseases of humans.

Acknowledgments

Appreciation is especially given to Robert Whitney, Jr., Joe Held, Benjamin Blood, Leo Whitehair, Dennis Johnsen and each of the participants and consultants of the task force, without whom this program would not have been possible. Many of the recollections of the task force meetings were derived from the excellent notes of Jo Fritz, for whom we are most grateful. The authors appreciate the critique of this manuscript by staff of the Comparative Medicine Program, National Center for Research Resources, NIH who have supported this program from the beginning and who continue to nurture its existence.

References

1 Interagency Primate Steering Committee (IPSC): National Primate Plan. NIH Publ No 80-1520. Bethesda, Interagency Primate Steering Committee, 1980.

2 Interagency Primate Steering Committee (IPSC): Report of the Ad Hoc Task Force to Develop a National Chimpanzee Breeding Program of the Interagency Research Primate Steering Committee. Tanglewood, June 1980.

3 Interagency Primate Steering Committee (IPSC): Report of the Task Force on the Use of the Need for Chimpanzees of the Interagency Primate Steering Committee. Bethesda, National Institutes of Health, 1978.

4 Johnsen DO: The need for using chimpanzees in research. Lab Animal 1987;16:19–23.

5 National Institutes of Health: A Program Report: NIH Chimpanzee Research Breeding Program. Bethesda, National Center for Research Resources, National Institutes of Health, 1990.

Thomas L. Wolfle, DVM, PhD, Institute of Laboratory Animal Research,
National Research Council, National Academy of Sciences,
2101 Constitution Avenue, N.W., Washington, DC 20418 (USA)

Eder G, Kaiser E, King FA (eds): The Role of the Chimpanzee in Research.
Symp, Vienna 1992. Basel, Karger, 1994, pp 87–90

Chimpanzee Demographics

Present Status and Future Outlook

J. Erwin

Diagnon Corporation, Rockville, Md., USA

Early Chimpanzee Research

Chimpanzees *(Pan troglodytes)* have been maintained in captivity for many years for purposes of scientific research and zoological exhibition. During the first half of the twentieth century, notable chimpanzee colonies were maintained in Europe and North America, especially for research on behavior and cognition. Examples include Madam Abreu's chimpanzee colony in Cuba, Koehler's facility in Tenerife [1], and Yerkes' program at Yale and Orange Park, Florida. Fundamental research on reproduction, sensation, neurobiology and physiology led to increased use in comparative medicine and eventually into space science programs.

Efforts to breed chimpanzees in captivity rather than removing them from the wild were promoted by Yerkes [2] and later by others but gained greater support as needs became apparent for a sustained chimpanzee resource for studies of immunology and infectious diseases. Fortunately, efforts had been under way for many years to breed chimpanzees in captivity [3]. The first chimpanzee birth in the New World was reported in 1915 from the colony in Cuba, and the first second-generation birth was reported by Yerkes in 1935.

Domestic Breeding

The signing by the USA and many other countries of the Convention on International Trade in Endangered Species (CITES) also indicated that importation of chimpanzees from countries of origin would cease to be a viable option. A breeding program was planned and was implemented in response to the AIDS crisis [4]. A part of the plan was the institution of a demographics program through the International Species Inventory System (ISIS) at the

Minnesota State Zoological Garden under the guidance of Dr. U.S. Seal and Dr. N. Flesness. Data are presented here from the ISIS demographics program along with data supplied directly from major chimpanzee colonies. In addition, ISIS data are included for the North American Species Survival Plan (SSP) of the Amerian Association of Zoological Parks and Aquariums (AAZPA). The total number of chimpanzees currently (mid-1992) on site at each of the following major research laboratories and breeding colonies are listed below:

(A)	Primate Foundation of Arizona	81
(B)	New Mexico Regional Primate Laboratory	337
(C)	White Sands Research Center	140
(D)	Southwest Foundation for Biomedical Research	222
(E)	University of Texas MD Anderson Science Park	151
(F)	New Iberia Research Center	247
(G)	Yerkes Regional Primate Research Center	205
(H)	LEMSIP, New York University	247
(I)	Bioqual Inc., Maryland	25

The total number at these institutions is 1,655 chimpanzees. These data and information on chimpanzees on loan to other settings suggest that there are about 1,750 chimpanzees in the North American research population. Unfortunately, not all research chimpanzees are included in the ISIS database.

The ISIS records on AAZPA/SSP participants include 214 (328) chimpanzees in zoological parks in North America. The total North American population maintained for research, breeding and zoological exhibition probably does not exceed 2,100 chimpanzees. As shown, about 650 chimpanzees outside North America are listed in the ISIS database.

Of the 9 research institutions listed above, 8 have produced a total of 99 births in 1990 and 103 births in 1991. No breeding occurs at Bioqual because the chimpanzees housed there are not sexually mature. Birth data for the 8 major breeding facilities follow:

Facility		1990 births (♂.♀)	1991 births (♂.♀)
(A)	PFA*	9 (4.5)	6 (5.1)
(B)	NMSUPL*	19 (13.6)	20 (11.9)
(C)	WSRC	11 (5.6)	8 (3.5)
(D)	SFBR	9 (5.4)	14 (6.8)
(E)	UTMDASP*	14 (7.7)	8 (5.3)
(F)	NIRC*	15 (10.5)	25 (11.14)
(G)	YRPRC*	12	10
(H)	LEMSIP	10	12

Institutional participants in the NIH Breeding and Research Program are designated by an asterisk. The male:female ratio for the institutions that provided gender data was 85:73 (a 1.16 bias toward males). The available ISIS data are not as current as those given above.

The proportion of captive chimpanzees of each of the three subspecific populations that occur in nature is unknown, but there is apparently a preponderance of *Pan troglodytes verus* from West Africa [5]. The extent to which captive breeding has resulted in subspecific crosses is unknown, but genetic studies in progress may answer this question. The ISIS data include inbreeding coefficients that can be used to maintain adequate genetic diversity in the captive population. The best evidence available indicate that the captive population will meet the present and future needs of the research community without removal of chimpanzees from countries of origin.

The Chimpanzee Breeding and Research Program requires that infants be left with their mothers to be socially reared to maximize the prospects of adequate reproductive behavior and rearing of offspring. The ultimate success of the breeding program depends on the reproductive adequacy of captive-born chimpanzees. Social housing and other environmental enrichment contributes to the well-being of individual chimpanzees and enhances the viability of the captive population [6–8]. There is much to be gained by communication regarding demographic data from research, breeding, conservation and exhibition situations to determine which patterns of socialization, care and housing are the most effective.

The Future

As the research and breeding population ages, provisions must be made for continued humane care of the chimpanzees [9]. Retirement support programs have been devised and have been implemented to some extent on the basis of user fees. The opportunity exists for cross-sectional and longitudinal research on aging correlates and for the development of innovative approaches to housing over the long term.

References

1 Koehler W: The Mentality of Apes (translated from the second revised edition). New York, Harcourt, Brace & Co, 1925.
2 Yerkes RM: Chimpanzees: A Laboratory Colony. New Haven, Yale University Press, 1943.
3 Maple T: Great apes in captivity: The good, the bad, and the ugly; in Erwin J, Maple T, Mitchell G (eds): Captivity and Behavior: Primates in Breeding Colonies, Laboratories, and Zoos. New York, Van Nostrand Reinhold, 1979, pp 239–272.

4 Blood B, Whitney R, Wolfle T: Development of the NIH Chimpanzee Breeding and Research Program; in Erwin J, Landon JC (eds): Chimpanzee Conservation and Public Health: Environments for the Future. Rockville, Diagnon/Bioqual, 1992, pp 33–38.
5 Erwin J: An introduction to chimpanzees and their environments; in Erwin J, Landon JC (eds): Chimpanzee Conservation and Public Health: Environments for the Future. Rockville, Diagnon/Bioqual, 1992, pp 1–5.
6 Metz B, Weld K, Erwin J: A system for documentation and assessment of behavioral responses of chimpanzees to environmental enrichment. Am J Primatol 1991;24:123.
7 Erwin J, Landon J, Bradbury R: Spacious quarantine suites for chimpanzees: Biological containment, management flexibility, and enhancement of species-typical behavior. Am J Primatol 1991; 24:98.
8 Erwin J, Landon J: Spacious biocontainment suites for chimpanzees in infectious disease research; in Erwin J, Landon J (eds): Chimpanzee Conservation and Public Health: Environments for the Future. Rockville, Diagnon/Bioqual, 1992, pp 65–69.
9 Landon J, Erwin J: A consensus statement regarding chimpanzee conservation and public health; in Erwin J, Landon J (eds): Chimpanzee Conservation and Public Health: Environments for the Future. Rockville, Diagnon/Bioqual, 1992.

J. Erwin, PhD, Diagnon Corporation, Rockville, MD 20850–3300 (USA)

Eder G, Kaiser E, King FA (eds): The Role of the Chimpanzee in Research.
Symp, Vienna 1992. Basel, Karger, 1994, pp 91–107

Reproduction in Common Chimpanzees with Reference to Problems of Fertility and Infertility

Kenneth G. Gould, Jeremy F. Dahl

Yerkes Regional Primate Research Center, Emory University, Atlanta, Ga., USA

Introduction

The common chimpanzee *(Pan troglodytes)* is one of the Great Apes (Hominoidea). *Pan* and *Homo* resemble each other in many ways, reflected in demonstrated similarities in physiology [1], endocrinology (both normal and abnormal) [2–4], immunology [5–8] and reproduction [2, 8, 9]. The study of reproduction in the chimpanzee is significant for a number of reasons. First, it is necessary to ensure the optimal husbandry and maintenance of the captive population. Second, it provides information needed for the responsible use of the species in understanding present medical problems such as AIDS [10–12]. Third, such studies provide information relevant to human reproduction and reproductive behavior [1, 13–16]. The focus of this paper is on the first of these, with secondary reference to the others.

Reproduction in this species has been studied in the captive population for over 60 years, with initial results available more than 50 years ago [17–19]. In retrospect, those studies clearly demonstrated resemblances in reproductive behavior and physiology between the chimpanzee and human which provided information regarding the time of ovulation and associated endocrine changes in the menstrual cycle that preceded similar data in the human. The studies of Yerkes and Elder, together with their colleagues, showed that the duration of the menstrual cycle of the chimpanzee female was slightly longer than that of the woman [17, 20]. Such observations were facilitated by the presence of a significant perineal swelling in the female which appeared to be a sensitive external indicator of the progress of the ovarian cycle [17]. Offspring born from matings at what were then known as the Yale Laboratories for Primate Biology, located in Orange Park, Florida, have been studied, since that time, as part of ongoing research. Gamma, the oldest captive chimpanzee, born at the Yale Laboratories for Primate Biology, died of old age in March 1992 at the Yerkes Primate Center in Atlanta. She was 59 years old.

The possible relationships between early learning and maternal influences on subsequent reproductive and maternal success are currently being studied at the Yerkes Center by Dr. Bard and her colleagues, as described in another paper in this symposium.

Male Reproduction

There has been an increased interest in the reproductive physiology and endocrinology of the male chimpanzee in the last decade, stimulated in part by increased emphasis on the responsible use of the chimpanzee as a model for biomedical research. Comparative studies of male endocrine maturation have been of special interest with regard to the occurrence of adrenarche [21]. It appears that, of the primate species thus far studied, only the chimpanzee shows this change in endocrine activity of the adrenal gland at the time of adolescence in a manner consonant with that in man [22–24]. In addition, more information on the endocrine status and pituitary responsiveness of the male chimpanzee has been obtained as part of several efforts. One is concerned with maximizing our understanding of the species for the development of conservation programs, and others are associated with the collection of basic information and use of the species in biomedical investigation related to human development and the study of infectious diseases, such as HIV [25, 26]. Information is now available regarding the changes in endocrine levels associated with age in the chimpanzee [21, 26–28]. The pattern of hormone changes during puberty has been described [22–24] and found to parallel those of the human to a large extent. Overall values for steroid and pituitary hormones are shown in figure 1. The chimpanzee responds to challenge with GnRH and N-methyl-*D*-aspartic acid in a manner analogous to the human male, although there may be some differences in sensitivity of the adult chimpanzee male to GnRH challenge [26].

In studies relevant to understanding fertility of the chimpanzee male, semen parameters of the pubertal and adult male have been documented [9, 27, 29–33], and these data are summarized in table 1. The chimpanzee male has a large testicle/body weight ratio ($\approx$0.27) [36] similar to that of other primate species with multimale breeding systems and is the largest for any ape genus. It has been argued that the relatively large testicular size is ultimately a result of a need for high levels of sperm production during intermale competition [36, 37].

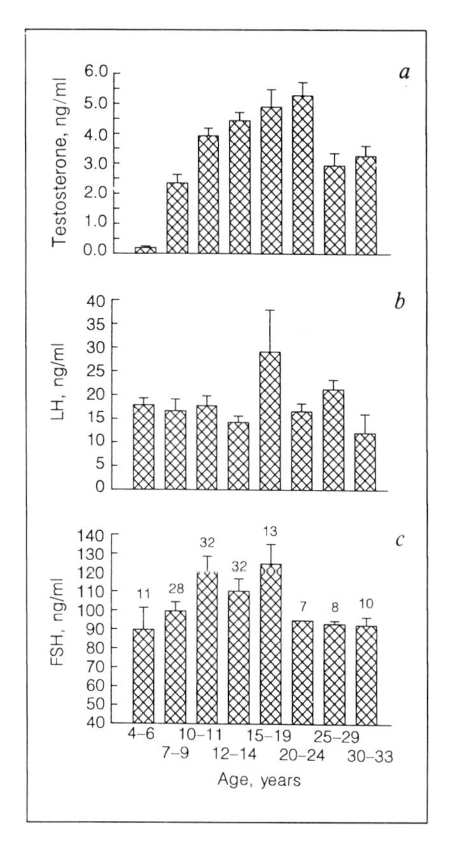

Fig. 1. Change with age, of testosterone *(a)*, LH *(b)* and FSH *(c)* in the serum of the common chimpanzee male. The figure above each bar is the sample size (data reproduced from Young et al. [26], with permission).

Table 1. Semen parameters of the common chimpanzee *(P. troglodytes)*

Method	Volume	Count, $\times 10^{-6}$/ml	Live, %	Motile, %	n	Reference No.
AV	2.4	620	70.35	NA	15	34
AV	3.4	2,561	80	76	25	27
Mast.	3.4	784	68	55	68[1]	31
Mast.	3.78	540	65	NA	27	34
RPE	1.0	280	84.4	NA	9	34
RPE	1.9	743	52.7	NA	11	35
RPE	2.5	1,000	89	85	6	Gould, unpubl.

AV = Artificial vagina; mast. = masturbated; RPE = rectal probe electrostimulation; NA = not available.

[1] Multiple, frequent ejaculates.

Fig. 2. Scanning electron micrograph of normal chimpanzee spermatozoa. Picture width = 12 μm.

However, preliminary analysis shows the chimpanzee to be similar to man in that the pattern of spermatogenesis is irregular [Hess R.A., pers. commun.] but dissimilar in that the sperm population shows few abnormal forms. The characteristic sperm morphology is shown in figure 2 [38, 39].

In this respect, the gorilla is more similar to man than is the chimpanzee, with a high percentage of morphologically abnormal spermatozoa in a 'fertile' ejaculate [38]. Further, the histology of the chimpanzee testis suggests that sperm production, expressed per unit of testis volume, appears lower than that of other species as demonstrated by the routine absence of spermatozoa from the lumen of the seminiferous tubule, even when the samples were obtained from males of proven fertility (fig. 3). The possible difference between these observations and the results that would be obtained by study of material from wild males remains to be investigated.

From the perspective of husbandry and breeding, there is little evidence of infertility, as opposed to nonproductivity, in the male chimpanzee [40]. When breeding success has been lowered as a result of reduced male 'fertility', this has been most often due to social factors and not to a reduction in fertilizing potential as identified by semen collection and analysis. According to the birth records of the captive population at the Yerkes Center there is no significant seasonality in births and thus no evidence for a 'breeding season' analogous to

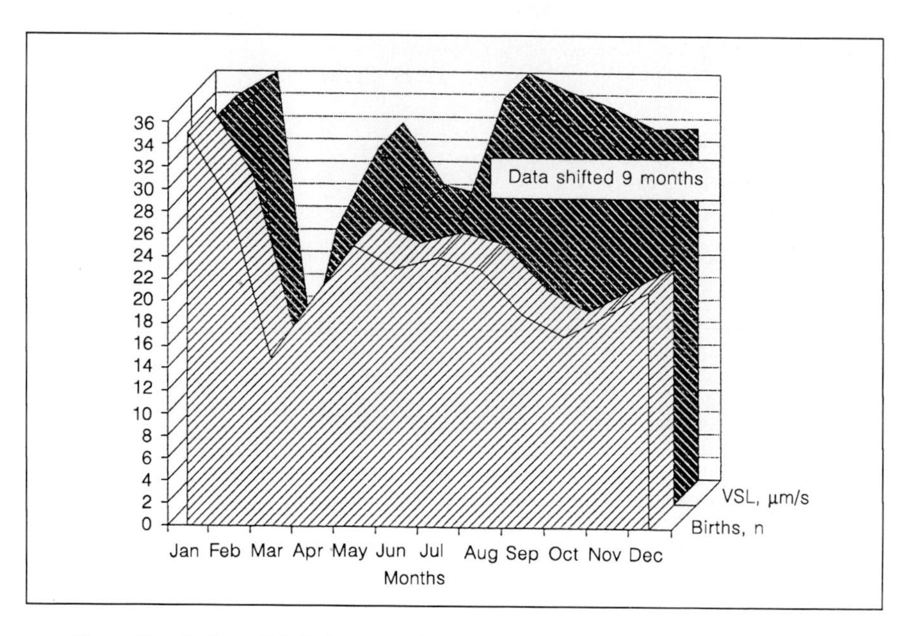

Fig. 4. Correlation of birth frequency in the Yerkes colony with annual change in sperm straight-line velocity (VSL). The data for VSL have been shifted 250 days to reflect the gestational length of the chimpanzee. Data for VSL were recovered recently. Data for birth frequency were recovered from historical and current records of the Yerkes colony (1930–1992, total=270).

and programs using a VP110 processor [51]. The reproducibility by CAMA of sperm motion parameters is dependent upon the baseline parameters, provided by the investigator with regard to recorder calibration, field sampling size and scale, and upon limiting values for such parameters as velocity and path linearity [53, 54, 60]. Inclusion of these parameters is critical when translating information from one species to another and from one laboratory to another. Information from CAMA is recovered from analysis of video recordings of sperm motion made immediately (<30 min) after collection or after liquefaction of the semen sample (2–4 h after collection). When used for sperm penetration assay analysis the video recording is made after overnight incubation at 4 °C [61]. Analysis in our laboratory is made using CAMA according to the parameters shown in table 2. Such analyses can provide important baseline data related to potential use of the species in development and evaluation of contraceptive agents. While there are similarities in the motion parameters of the chimpanzee and human, it appears that significant differences exist between the macaque, chimpanzee and man with regard to such characters as straight-line velociy and linearity (fig. 5). It is now possible to successfully freeze sperm samples from the chimpanzee and use

Table 2. Parameters used for collection and calculation of CAMA data from great apes

Parameter	Value
Frame rate, frames/s	30
Duration of data capture, frames	90
Minimum path length, frames	30
Minimum motile speed, μm/s	10
Maximum burst speed, μm/s	350
Distance scale factor, μm/pixel	1.3880
Camera aspect ratio	1.0590
ALH path smoothing factor,[1] frames	7
Cent. X search neighborhood,[2] pixels	4
Cent. Y search neighborhood,[2] pixels	2
Cent. cell size minimum,[3] pixels	2
Cent. cell size maximum,[3] pixels	10
Path maximum interpolation,[4] frames	5
Path prediction percentage,[5] %	0
Depth of sample, μm	10
Video processor model	VP110

[1] Factor used to reduce the effect of random movement on calculation of lateral head displacement (ALH).
[2] Area of the frame (in pixels) to be checked as part of the area into which the subject could move from one frame to the next.
[3] Definition of the subject size to be considered as a sperm head.
[4] Number of frames to be used to account for missing data, due to sperm swimming 'out of focus'.
[5] This value accounts for the predictability of the sperm path. As sperm do not travel in a predictable direction, this value is low, here set at zero.

those samples, when thawed, to initiate pregnancy [51]. Quantitative motion analysis has the potential to facilitate such conservation efforts by aiding in the identification of those samples best suited for freeze preservation.

The chimpanzee male has been shown to be susceptible to diseases which, in man, can be associated with impaired reproductive function. At this time, however, the species has been actively used only in limited study of such diseases or their control [40, 62]. It has been shown that the male chimpanzee is susceptible to experimental infection with *Mycoplasma genitalium*, with the induction of an obvious genital tract infection and shedding of the organism [62]. While spontaneous infection with ureaplasmas and mycoplasmas has been demonstrated by recovery of the organisms from the genital tract of approximately one third of males evaluated, as in man it has not been possible to

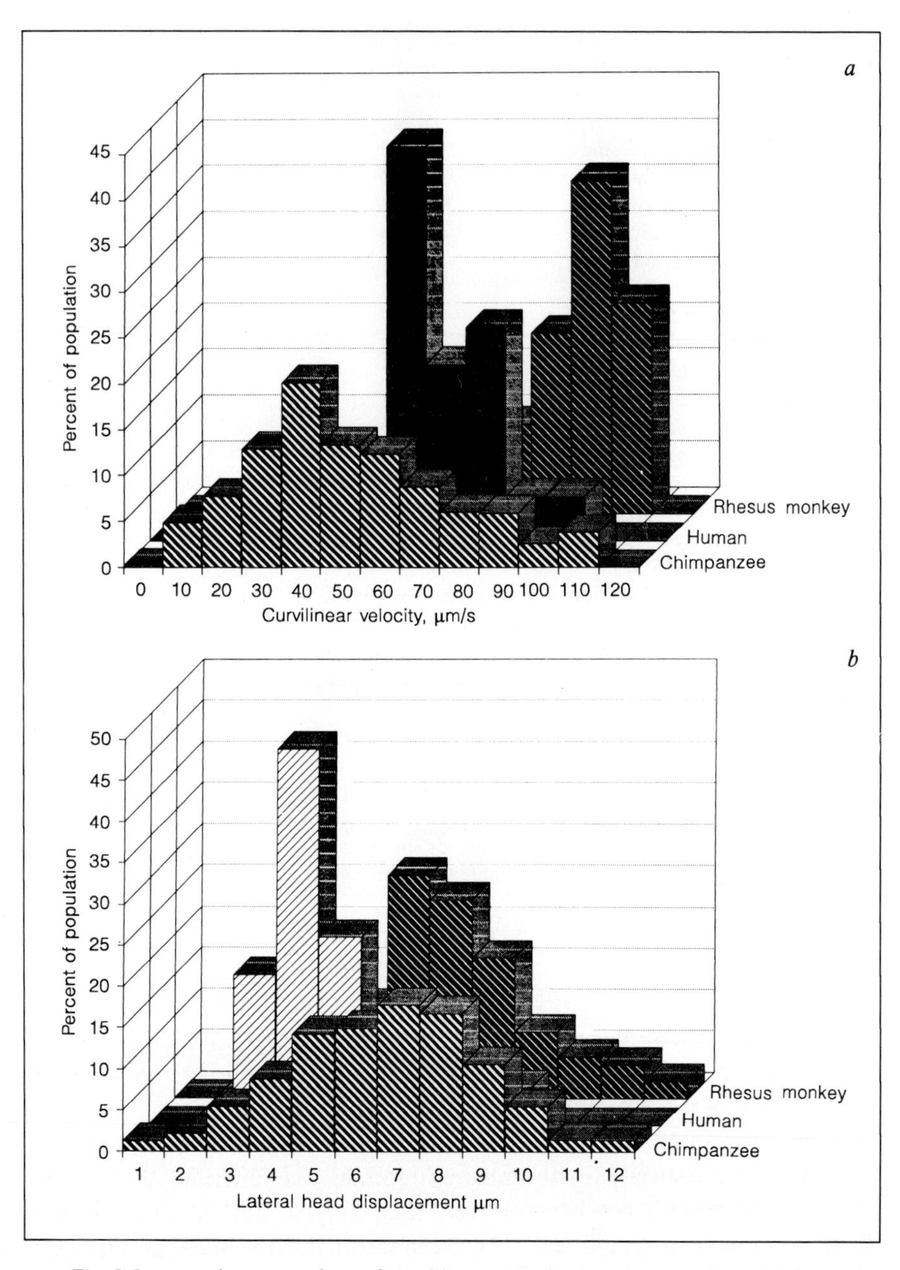

Fig. 5. Interspecies comparison of curvilinear velocity *(a)* and lateral head displacement *(b)* of sperm analyzed using CAMA. A difference in the distribution of velocity for the three species is clear. The histogram is based upon analysis of at least 2,500 sperm for each species.

correlate those isolations with infertility [40, 62]. The possible role of *Chlamydia* spp. has not been evaluated. *Herpes hominis* type II (human genital herpes) has been identified in the chimpanzee, but, as in man, the infection has not been correlated with infertility despite the presence of clear and characteristic lesions. There has been a single report of arteriosclerosis of the spermatic arteries of the chimpanzee associated with infertility, despite continued spermatogenesis. That association, however, is circumstantial [63].

Female Reproduction

It has been recognized for a long time that the female chimpanzee exhibits a menstrual cycle of approximately 35 days duration, with an endocrine pattern similar to that of the woman [18, 19]. The chronology of early publications of researchers such as Bingham, Tinklepaugh, Yerkes and Elder has been documented previously [64]. In 1936 Yerkes and Elder systematically described the changes in perineal swelling of the chimpanzee during the menstrual cycle [17, 19]. Subsequently, and with increasing precision as techniques have improved, the basic endocrine pattern of the normal menstrual cycle and its correlation to physical events of perineal tumescence and ovulation have been documented [1, 65]. The metabolism of the steroid hormones in the female chimpanzee has been elucidated, and found to parallel, for the most part, that of the woman. That the perineal swelling is directly under the control of estrogen was demonstrated by the detection and quantification of estrogen receptors in the perineal tissue [66] and by reduction of the swelling in the presence of estrogen receptor antagonist [67, 68]. Graham [64] demonstrated the inhibitory effect of progesterone on the induction of the perineal swelling and documented the presence of spontaneous predecidualization in the normal menstrual cycle, a phenomenon shared with the woman and absent in the macaque. Those observations provide a basis for current investigations into the modification of the luteal phase of the pregnant cycle to enhance the rate of initiation of pregnancy. Such studies have application to the human female. Using this method two patterns of perineal swelling during the normal menstrual cycle have shown two types of regular cycle [69], one with a short preswollen phase, a long swollen phase and a relatively long postswollen phase (SLL), and others with a long preswollen phase and shortened swollen and postswollen phases (LSS). In the SLL type, the preovulatory LH peak can be predicted 15 days in advance, and the subsequent luteal-phase progesterone levels are normal. The LSS type, however, exhibits a variability in the time of the LH peak and midluteal pregnanediol levels approximately half those of the SLL type (fig. 6). While the alteration in luteal hormone patterns is not sufficient to permit

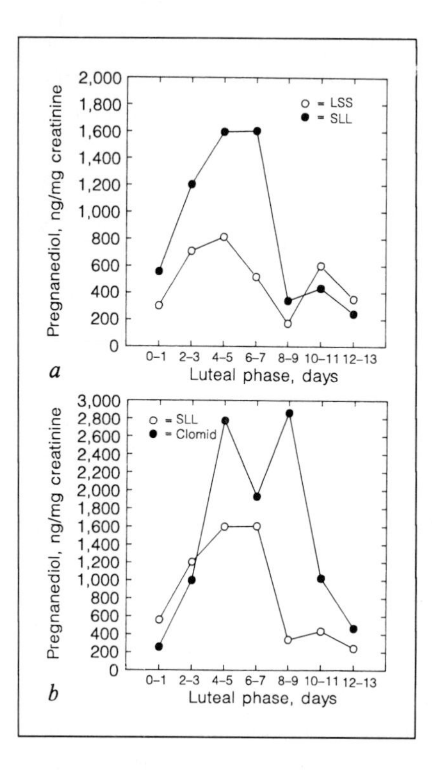

Fig. 6. a Comparison of luteal-phase levels of urinary pregnanediol for two types of swelling cycle (LSS and SLL) described in the text. *b* Comparison of urinary pregnanediol levels for a normal SLL cycle and for a cycle treated with clomiphene citrate (Clomid).

classification as 'inadequate luteal phase', a significantly higher rate of successful artificial inseminations has been achieved in SLL cycles or hormonally synchronized cycles which exhibited the SLL pattern. In addition, natural pregnancy has been observed to occur in females at the time of an SLL pattern cycle while more than 12 months of LSS cycle pattern (in the presence of fertile males) had not been associated with pregnancy.

Not surprisingly, observations regarding menstrual cyclicity and breeding patterns obtained from the captive population demonstrate similarities to those derived from observation of the wild population [70–72]. In general, captive females mature at an earlier age, although probably at an equivalent body weight to their wild relations [70, 73].

The average age at first pregnancy is approximately 9 years for the captive population but 2–3 years later in the wild [70, 72, 73]. There appears to be a period of adolescent sterility similar to that seen in the woman which lasts for 6–18 months. In contrast to the woman, the aged common chimpanzee *(P. troglodytes)* does not show a cessation of reproductive function equivalent to menopause [74, 75], although this may not be so for the pygmy chimpanzee *(P. paniscus).* Menstrual cyclicity continues into the fourth and fifth decades of life although cycle frequency is decreased

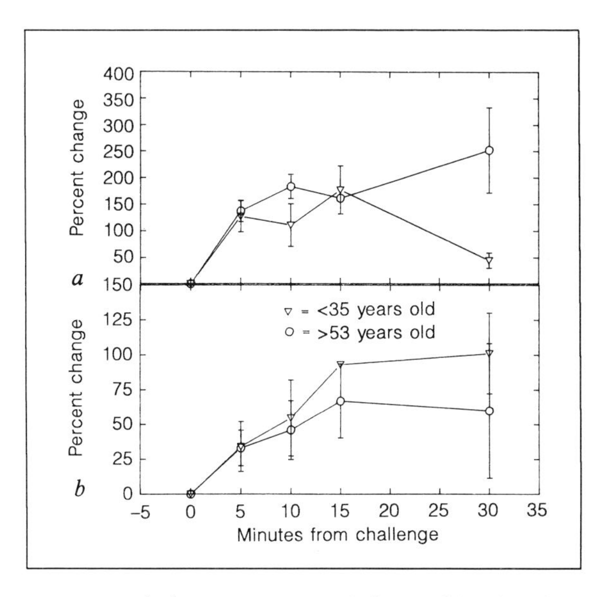

Fig. 7. Pituitary response to a challenge with 100 μg GnRH administered as an intravenous bolus. The figure compares the response of LH *(a)* and FSH *(b)* for a group of females less than 35 years of age with that of a group of females greater than 53 years of age (n = 3).

from 9.4/year for animals <25 years old to 8.5/year for animals >35 years old. The pregnancy rate for apparently normal, sexually active cycles is reduced from 20% in the younger animals to less than 4% in the aged group [75]. Further, menopause, as documented by alteration in pituitary response to GnRH challenge as well as alteration or extinction of menstrual cyclicity, has not been demonstrated, even in the oldest females of record (fig. 7).

Patterns of perineal swelling in the luteal phase (fig. 8) are similar to patterns of edema clearly associated with premenstrual syndrome in women [76]. Edema is the most frequent symptom clustered with premenstrual syndrome within a wide range of other changes tested [77]. Further evaluation of the chimpanzee female for study of the endocrine correlates of premenstrual syndrome is clearly indicated [76].

Infertility in the female chimpanzee has been associated with endocrinological factors such as inadequate luteal phase [78], amenorrhea as a result of various causes including lactation and physical factors such as oviductal obstruction and the presence of pelvic adhesions. Endometriosis has not been identified as contributing to infertility in the chimpanzee, although there is no reason to believe that it cannot occur. Other pathological situations such as membranous dysmenorrhea (endometrial casting) have also been recorded [79, 80]. Although it has been suggested that this is associated with the loss of an extrauterine pregnancy [81], this does not appear to be the case in those

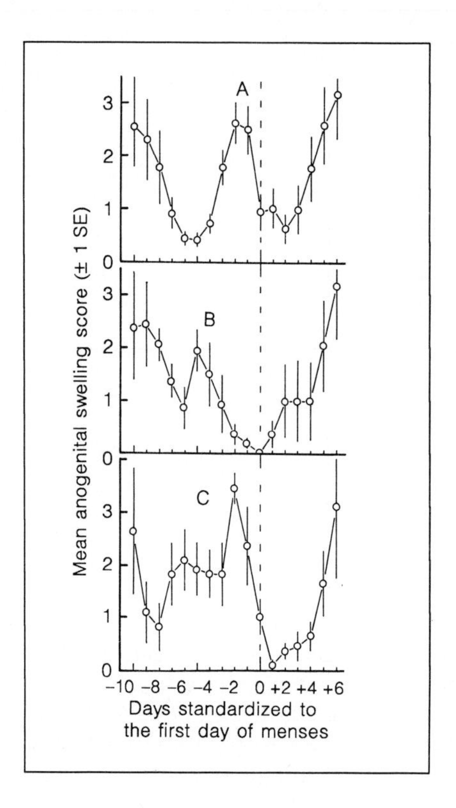

Fig. 8. Variation in the mean anogenital swelling score demonstrating the three patterns of swelling identified in the chimpanzee (n=27, 16, 11 for types A, B and C, respectively).

instances studied in the chimpanzee. The etiology thus remains unknown. However, although the fertility/parity of the females described from the colony of the TNO, the Netherlands organization for applied scientific research [80], was not given, females in the Yerkes colony which have demonstrated endometrial casting have subsequently conceived and delivered normal infants [79]. Abortion in the gorilla has been associated with infection by *Shigella flexneri*, but this has not been demonstrated for the chimpanzee [82]. It has been shown that the female chimpanzee is susceptible to experimental infection with *M. genitalium*, with the induction of an obvious genital tract infection and shedding of the organism [62]. While spontaneous infection with ureaplasmas and mycoplasmas has been demonstrated by recovery of the organisms from the genital tract of approximately two thirds of females evaluated, as in women it has not been possible to correlate those isolations with infertility [40, 62]. The possible role of *Chlamydia* spp. has not been evaluated.

An increased interest in investigation of the mechanisms responsible for control of amenorrhea in the chimpanzee is the result of a possible association, as in the woman, (a) with physical and/or social stress, (b) with pregnancy and

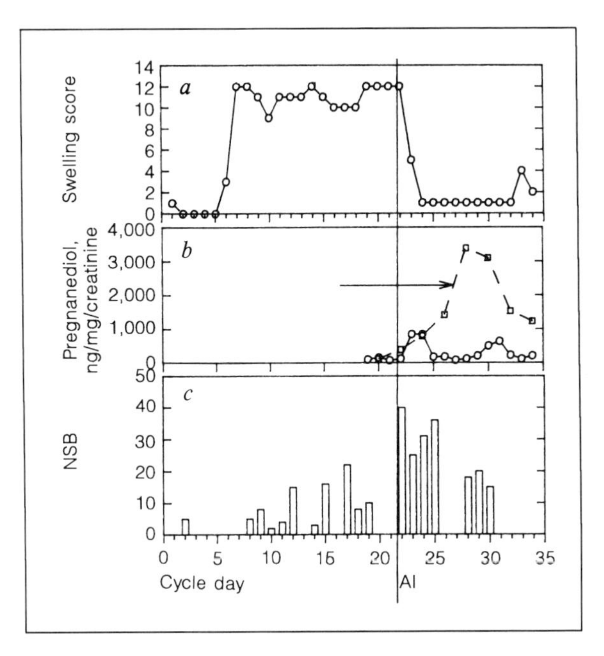

Fig. 9. Demonstration of the correlation of perineal swelling *(a)*, luteal-phase pregnanediol level *(b)* and nipple autostimulation behaviour (NSB, *c*) in a female chimpanzee (Marietta). The level of pregnanediol associated with the normal menstrual cycle (□) is provided for comparison. AI = Artificial insemination.

(c) with lactation. While husbandry requirements prompt the investigation of means to reduce the duration of lactational amenorrhea in the chimpanzee, elucidation of the endocrine bases for this amenorrhea will also be of potential value in the management of 'stress-related' amenorrhea in women. Behaviors exhibited by chimpanzees such as nipple autostimulation appear to be correlated with altered endocrine patterns (fig. 9) and impaired pregnancy rate, as opposed to low fertilization rate, subsequent to artificial insemination. Nipple stimulation behaviors have been implicated as contributing to prolonged postpartum amenorrhea [83, 84], and this can be terminated by treatment with a dopamine receptor agonist. However, recent studies in our laboratory show that treatment with a dopamine receptor agonist does not terminate nipple stimulation behaviors [16]. It appears that continuation of this pattern of behavior may cause lowered luteal-phase progesterone production or some other disruption of the luteal phase that interferes with implantation and the establishment of pregnancy. High oxytocin concentrations [85, 86], rather than elevated prolactin levels [86], may be a cause of this lowering of the pregnancy rate, a hypothesis currently under test at the Yerkes Center [16].

Conclusion

There is an increasing volume of information available regarding the normal reproductive parameters of the common chimpanzee. In general these data demonstrate similarities between the chimpanzee and man but reveal some characters that differ significantly from other primate species studied. Such differences, for the male, can be related to the breeding strategy of the chimpanzee. Overall, there is little evidence for disease-induced infertility in the captive population. This suggests that current maintenance and medical programs are effective with regard to potential expansion of the captive population and that population growth can be controlled by: (1) careful and appropriate monitoring of female cyclicity, (2) application of appropriate treatment protocols for subjects with lowered fertility and (3) manipulation of social groupings by selective movement of either males or females and development of improved rearing methods.

Ongoing research efforts need to be maintained with special emphasis on identification, monitoring and treatment of individuals, both male and female, with a particularly valuable genome and with regard to the development of improved methods for gamete collection and banking related to the potential for collection of genetic material from the endangered wild population.

Acknowledgements

This work was supported by NIH grants #RR-00165 to the Yerkes Primate Center, and HD-26076 and RR-03587 to Dr. K. Gould. The technical and secretarial assistance of Mss. S. Phythyon, J. Johnson and J. Lovell is gratefully acknowledged.

References

1 Graham CE: The chimpanzee: A unique model for human reproduction; in Antikatzides Th, Erichsen S, Spiegel A (eds): The Laboratory Animal in the Study of Reproduction: 6th ICLA Symposium, Thessaloniki 1975. Stuttgart, Fischer, 1976, pp 29–38.

2 Graham CE, Gould KG, Collins DC, Preedy JRK: Regulation of gonadotropin release by luteinizing hormone-releasing hormone and estrogen in chimpanzees. Endocrinol 1979;105:269–275.

3 Winter JSD, Faiman C, Hobson WC, Prasad AV, Reyes FI: Pituitary-gonadal relations in infancy. I. Patterns of serum gonadotropin concentrations from birth to four years of age in man and chimpanzee. J Clin Endocrinol Metab 1975;40:545–551.

4 Winterer J, Merriam GR, Gross E, et al: Idiopathic precocious puberty in the chimpanzee: A case report. J Med Primatol 1984;13:73–79.

5 Bontrop RE, Broos LAM, Pham K, Bakas RM, Otting N, Jonker M: The chimpanzee major histocompatibility complex class II DR subregion contains an unexpectedly high number of beta-chain genes. Immunogenetics 1990;32:272–280.

6 Haas GG, Nahhas F: Failure to identify HLA, ABC and Dr antigens on human sperm. Am J Reprod Immunol Microbiol 1986;10:39–46.

7 Murayama Y, Fukao K, Noguchi A, Takenaka O: Epitope expression on primate lymphocyte surface antigens. J Med Primatol 1986;15:215–266.

8 Socha WW: Blood groups as genetic markers in chimpanzees: Their importance for the national chimpanzee breeding program. Am J Primatol 1981;1:3–13.
9 Martin DE, Gould KG: The male ape genital tract and its secretions; in Graham CE (ed): Reproductive Biology of the Great Apes: Comparative and Biomedical Perspectives. New York, Academic Press, 1981, pp 127–162.
10 Kolbe H, Kasumi K, Chaput A, Reinhart T, Muchmore E: Immunization of chimpanzees confers protection against challenge with human immunodeficiency virus. Proc Natl Acad Sci USA 1991; 88:542–546.
11 Peeters M, Honoré C, Huet T, et al: Isolation and partial characterization of an HIV-related virus occurring naturally in chimpanzees in Gabon. AIDS 1989;3:625–630.
12 Van Eendenburg JP, Yagello M, Girard M, et al: Cell-mediated immune proliferative responses to HIV-1 of chimpanzees vaccinated with different vaccinia recombinant viruses. Aids Res Hum Retroviruses 1989;5:41–50.
13 Gibbons A: Our chimp cousins get that much closer. Science 1990;250:376.
14 Nadler RD, Dahl JF, Collins DC, Gould KG, Wilson DC: Effects of oral contraceptives on chimpanzees: A preliminary report; in Eley RM (ed): Comparative Reproduction in Mammals and Man: Proceedings of a Conference of the National Centre for Research in Reproduction. Nairobi, Institute of Primate Research, National Museums of Kenya, 1987, pp 30–33.
15 Nadler RD, Dahl JF, Gould KG, Collins DC: Effects of an oral contraceptive on sexual behavior of chimpanzees *(Pan troglodytes)*. Arch Sex Behav, in press.
16 Nadler RD, Dahl JF, Gould KG, Collins DC: Hormone levels and ano-genital swelling of female chimpanzees as a function of estrogen dosage in a combined oral contraceptive. Proc Soc Exp Biol Med 1992;201:73–79.
17 Elder JH, Yerkes RM: The sexual cycle of the chimpanzee. Anat Rec 1936;76:119–143.
18 Yerkes RM: Sexual behavior in the chimpanzee. Hum Biol 1939;11:78–111.
19 Yerkes RM, Elder JH: Oestrus, receptivity and mating in the chimpanzee. Comp Psychol Monogr 1936;13:1–39.
20 Young WC, Yerkes RM: Factors influencing the reproductive cycle in the chimpanzee: The period of adolescent sterility and related problems. Endocrinology 1943;33:121–154.
21 Martin DE, Swenson RB, Collins DC: Correlation of serum testosterone levels with age in male chimpanzees. Steroids 1977;29:471–481.
22 Hobson WC, Fuller GB, Winter JSD, Faiman C, Reyes FI: Reproductive and endocrine development in the great apes; in Graham CE (ed): Reproductive Biology of the Great Apes. New York, Academic Press, 1981, pp 83–103.
23 Nadler RD, Wallis J, Roth-Meyer C, Cooper RW, Baulieu EE: Hormones and behavior of prepubertal and peripubertal chimpanzees. Horm Behav 1987;21:118–131.
24 Oerter KE, Uriarte MM, Rose SR, Barnes KM, Cutler GB: Gonadotropin secretory dynamics during puberty in normal girls and boys. J Clin Endocrinol Metab 1990;71:1251–1258.
25 Moor-Jankowski J: The first successful AIDS vaccine: Immunization of chimpanzees confers protection against challenge with human immunodeficiency virus. J Med Primatol 1991;20:47–48.
26 Young LG, Gould KG, Smithwick EB: Selected endocrine parameters of the adult male chimpanzee. Am J Primatol, in press.
27 Gould KG, Young LG, Smithwick E, Phythyon S: Semen characteristics of the adult male chimpanzee *(Pan troglodytes)*. Am J Primatol 1992;29:221–232.
28 Marson J, Meuris S, Cooper RW, Jouannet P: Puberty in the male chimpanzee: Time-related variations in luteinizing hormone, follicle-stimulating hormone, and testosterone. Biol Reprod 1991;44:456–460.
29 Bavister BD: Oocyte maturation and in vitro fertilization in the rhesus monkey; in Stouffer RL (ed): The Primate Ovary. New York, Plenum Press, 1987, pp 119–137.
30 Gould KG, Warner H, Martin DE: Rectal probe electroejaculation of primates. J Med Primatol 1978;7:213–222.
31 Marson J, Gervais D, Meuris S, Cooper RW, Jouannet P: Influence of ejaculation frequency on semen characteristics in chimpanzees *(Pan troglodytes)*. J Reprod Fertil 1989;85:43–56.
32 Marson J, Meuris S, Cooper RW, Jouannet P: Puberty in the male chimpanzee: Progressive maturation of semen characteristics. Biol Reprod 1991;44:448–455.
33 Marson J, Meuris S, Moysan F, Gervais D, Cooper RW, Jouannet P: Cellular and biochemical characteristics of semen obtained from pubertal chimpanzees by masturbation. J Reprod Fertil 1988;82:199–207.

34 Gould KG: Ovulation detection and artificial insemination. Am J Primatol 1982;1:15–25.
35 Bader H: Electroejaculation in chimpanzees and gorillas and artificial insemination in chimpanzees. Zoo Biol 1983;2:307–314.
36 Harcourt AH, Harvey PH, Larson SG, Short RV: Testis weight, body weight and breeding system in primates. Nature 1981;293:55–57.
37 Moller AP: Ejaculate quality, testes size and sperm competition in primates. J Hum Evol 1988; 17:479–488.
38 Gould KG, Martin DE, Hafez ESE: Mammalian spermatozoa; in Hafez ESE (ed): SEM Atlas of Mammalian Reproduction. Tokyo, Igaku Shoin, 1975, pp 42–57.
39 Martin DE, Gould KG: Normal and abnormal hominoid spermatozoa. J Reprod Med 1975; 14:204–209.
40 Taylor-Robinson D, Barile MF, Furr PM, Graham CE: Ureaplasmas and mycoplasmas in chimpanzees of various breeding capacities. J Reprod Fertil 1987;81:169–173.
41 Gould KG, Martin DE: Artificial insemination of nonhuman primates; in Benirschke K (ed): Primates: The Road to Self-Sustaining Populations. New York, Springer, 1986, pp 425–443.
42 Warner H, Martin DE, Keeling ME: Electroejaculation of the great apes. Ann Biomed Eng 1974; 2:419–432.
43 Gould KG, Martin DE: Comparative morphology of primate spermatozoa using scanning electron microscopy. II. Families Cercopithecidae, Lorisidae, Lemuridae. J Hum Evol 1978;7:637–642.
44 Vandevoort CA, Neville LE, Tollner TL, Field LP: Non-invasive semen collection from adult orang-utans. Zoo Biol 1993;12:257–266.
45 Fussell EN, Franklin LW, Franta RC: Collection of chimpanzee sperm with an artificial vagina. Lab Anim Sci 1973;23:252–255.
46 Lanzendorf SE, Gliessman PM, Archibong AE, Alexander M, Wolf DP: Collection and quality of rhesus monkey semen. Mol Reprod Dev 1990;25:61–66.
47 Settlage DSF, Hendrickx AG: Electroejaculation technique in *Macaca mulatta*. Fertil Steril 1974; 25:157–159.
48 Roussel JD, Austin CR: Improved electroejaculation of primates. J Inst Anim Technol 1968; 19:22–32.
49 Check JH, Bollendorf AM, Press MA, Breen EM: Noninvasive techniques for improving fertility potential of retrograde ejaculates. Arch Androl 1990;25:271–276.
50 Gould KG, Mann DR: Comparison of electrostimulation methods for semen recovery in the rhesus monkey *(Macaca mulatta)*. J Med Primatol 1988;17:95–103.
51 Gould KG, Styperek RP: Improved methods for freeze preservation of chimpanzee sperm. Am J Primatol 1989;18:275–284.
52 Gould KG, Young LG: Acquisition of fertilizing capacity by chimpanzee sperm. Folia Primatol 1990;54:105–108.
53 Knuth UA, Nieschlag E: Comparison of computerized semen analysis with the conventional procedure in 322 patients. Fertil Steril 1988;49:881–885.
54 Mack SO, Wolf DP, Tash JS: Quantitation of specific parameters of motility in large numbers of human sperm by digital image processing. Biol Reprod 1988;38:270–281.
55 Young LG, Gould KG: Surface components of ejaculated chimpanzee sperm. Arch Androl 1982; 8:15–20.
56 Holt WV, Moore HDM, Hillier SG: Computer assisted measurement of sperm swimming speed in human semen: Correlation of results with in vitro fertilization assays. Fertil Steril 1985;44:112–119.
57 Katz DF, Overstreet JW: Sperm motility assessment by videomicrography. Fertil Steril 1981; 35:188–193.
58 Chan SYW, Wang C, Ng M, et al: Evaluation of computerized analysis of sperm movement characteristics and differential sperm tail swelling patterns in predicting human sperm in vitro fertilizing capacity. J Androl 1989;10:133–138.
59 Vantman D, Banks SM, Koukoulis G, Dennison L, Sherins RJ: Assessment of sperm motion characteristics from fertile and infertile men using a fully automated computer-assisted semen analyzer. Fertil Steril 1989;51:156–161.
60 Mack SO, Tash JS, Wolf DP: Effect of measurement conditions on quantification of hyperactivated human sperm subpopulations by digital image analysis. Biol Reprod 1989;40:1162–1169.
61 Bolanos JR, Overstreet JW, Katz DF: Human sperm penetration of zona-free hamster eggs after storage of the semen for 48 hours at 2 °C to 5 °C. Fertil Steril 1983;39:536–541.

62 Tully JG, Taylor-Robinson D, Rose DL, Furr PM, Graham CE, Barile MF: Urogenital challenge of primate species with *Mycoplasma genitalium* and characteristics of infection induced in chimpanzees. J Infect Dis 1986;153:1046–1054.
63 Pearson GR, Slinger WB: Arteriosclerosis of the spermatic arteries of a chimpanzee *(Pan troglodytes)*. Vet Pathol 1982;19:710–712.
64 Graham CE: A survey of advances in chimpanzee reproduction; in Bourne GH (ed): Progress in Ape Research. New York, Academic Press, 1977, pp 177–190.
65 Graham CE, Gould KG, Wright K, Collins DC: Luteal estrogen secretion and decidualization in the chimpanzee; in Chivers DJ, Ford EHR (eds): Recent Advances in Primatology. New York, Academic Press, 1978, vol IV, pp 209–211.
66 Ozasa H, Gould KG: Demonstration and characterization of estrogen receptor in chimpanzee sex skin: Correlation between nuclear receptor levels and degree of swelling. Endocrinology 1982; 111:125–131.
67 Gould KG, Martin DE, Warner H: Improved method for artificial insemination in the great apes. Am J Primatol 1985;8:61–65.
68 Lefevre B, Gougeon A, Peronny H, Testart J: A gonadotropin-releasing hormone agonist and an activator of protein kinase C improve in vitro oocyte maturation in *Macaca fascicularis*. Gamete Res 1988;21:193–197.
69 Dahl JF, Nadler RD, Collins DC: Monitoring the ovarian cycles of *Pan troglodytes* and *P. paniscus*: A comparative approach. Am J Primatol 1991;24:195–209.
70 Coe CL, Connolly AC, Kraemer HC, Levine S: Reproductive development and behavior of captive female chimpanzees. Primates 1979;20:571–582.
71 Goodall J: The behaviour of free-living chimpanzees in the Gombe stream reserve. Anim Behav Monogr 1968;1:165–311.
72 Tutin CEG: Sexual Behavior and Mating Patterns in a Community of Wild Chimpanzees *(Pan troglodytes schweinfurthii)*; PhD diss, 1975.
73 Clark CB: A preliminary report on weaning among chimpanzees of the Gombe national park, Tanzania; in Chavalier-Skolnikoff S, Poirier FE (eds): Primate Bio-Social Development. New York, Garland, 1977, pp 235–260.
74 Gould KG, Flint M, Graham CE: Chimpanzee reproductive senescence: A possible model for evolution of the menopause. Maturitas 1981;3:157–166.
75 Graham CE: Reproductive function in aged female chimpanzees. Am J Phys Anthropol 1979;50:291–300.
76 Dahl JF, Gould KG, Phythyon SP, Wosar M: Premenstrual edema of the sex swelling in chimpanzees. In press.
77 Magos AL, Brincat M, Studd JWW: Trend analysis of the symptoms of 150 women with a history of the premenstrual syndrome. Am J Obstet Gynecol 1986;155:277–282.
78 Fritz MA: Inadequate luteal function and recurrent abortion: Diagnosis and treatment of luteal phase inadequacy. Semin Reprod Endocrinol 1988;6:129–143.
79 Gould KG, Martin DE: The female ape genital tract and its secretions; in Graham CE (ed): Reproductive Biology of the Great Apes: Comparative and Biomedical Perspectives. New York, Academic Press, 1981, pp 105–125.
80 Solleveld HA, Van Zwieten MJ: Membranous dysmenorrhea in the chimpanzee *(Pan troglodytes)*: A report of four cases. J Med Primatol 1978;7:19–25.
81 Novak ER, Woodruff JD: Novak's Gynecologic and Obstetric Pathology. Philadelphia, Saunders, 1974.
82 Swenson RB, McClure HM: Septic abortion in a gorilla due to *Shigella flexneri*. Annu Proc Am Assoc Zool Vet 1974, pp 195–196.
83 Graham CE, Struthers EJ, Hobson WC, Faiman C: Prolonged postpartum amenorrhea in chimpanzees: Treatment and etiology; in Eley RM (ed): Comparative Reproduction in Mammals and Man. Nairobi, National Museums of Kenya, 1989, pp 8–11.
84 Graham CE, Struthers EJ, Hobson WC, et al: Postpartum infertility in common chimpanzees. Am J Primatol 1991;24:245–255.
85 Luck MR: A function for ovarian oxytocin. J Endocrinol 1989;121:203–204.
86 Samson WK, Lumpkin MD, McCann SM: Evidence for a physiological role for oxytocin in the control of prolactin secretion. Endocrinol 1986;119:554–560.

K.G. Gould, PhD, MRCVS, Division of Reproductive Biology, Yerkes Regional Primate Research Center, Emory University, Altanta, GA 30322 (USA)

Eder G, Kaiser E, King FA (eds): The Role of the Chimpanzee in Research.
Symp, Vienna 1992. Basel, Karger, 1994, pp 108–120

The Emergence of the Chimpanzee in Research

Frederick A. King, Cathy J. Yarbrough

Yerkes Regional Primate Research Center, Emory University, Atlanta, Ga., USA

The Yerkes Primate Research Center, established 62 years ago by Dr. Robert M. Yerkes of Yale University, in many respects reflects the steps by which chimpanzees emerged in research and their pattern of use in biomedical and behavioral studies during this century. Although the Center was established in 1930, Dr. Yerkes' ideas and design for a primate research institute actually originated as early as 1900, while he was a graduate student in psychology at Harvard University. By 1924 he had founded a primate laboratory at Yale University in New Haven, Connecticut [1]. As Dr. Leonard Carmichael recalled in 1968: 'It suddenly became clear (to Yerkes) that what was needed to unlock the complex problems of human and animal behavior was a research institute for the comparative study of mammals and man, with special reference to the study of the great apes' [2].

Yerkes, a dedicated psychobiologist, regarded the study of behavior as a source of information about intelligence and conscious processes and a tool for the comparison of these processes across species. He reasoned that primates, because of their evolutionary closeness to humans, were the species whose behavior could shed most light upon the roots of human behavior, both social and cognitive [1, 3]. Yerkes was not alone, in these early years, in believing that primate studies would benefit our understanding of human behavior. Two eminent anthropologists, Drs. Earnest A. Hooten and Alfred L. Kroeber, supported primate behavior studies because the results could improve understanding of human culture and behavior. Hooten, a leading figure in the development of research on cultural anthropology, dedicated his 1942 book *Man's Poor Relations* to Dr. Yerkes [4, 5]. In the book, Hooten wrote that research on primate behavior may be as relevent to 'the student of man as does the investigation of the social life and psychology of contemporary savages' [5]. Kroeber, who was also impressed with Yerkes studies [4], wrote in 1928 that studies of primates, particularly chimpanzees, could be important to understanding the origin of culture [6]. Using such criteria as religion and ethics, Kroeber concluded that chimpanzees lacked culture but possessed 'reactions and faculties closely akin to our own and manifesting at least some measure of the basal psychic ingredient, which enters into culture' [6].

In the early decades of this century, primates were mainly limited to the stage, museums and zoos and were rarely studied scientifically. The first scientific study of chimpanzees in a zoo likely occurred at the New York Zoological Park [7]. While primates were frequently described in the popular and scientific press [8, 9], these discourses were rarely scientific. In 1924, Yerkes wrote: 'Where we most need reliable, systematic, detailed descriptions, we find observational fragments cemented together with guesses, some shrewd, some ridiculous' [10].

Through his establishment in 1930 of the Yale Laboratories of Primate Biology in Orange Park, Florida, now known as the Yerkes Regional Primate Research Center, located at Emory University in Atlanta since 1965, Yerkes set in motion the scientific studies that provided the reliable, systematic and detailed descriptions that he called for in 1924. As a result, in the field of behavioral primatology, Yerkes, who also introduced Pavlov to the USA [11,12], today stands as the founder of comparative behavioral studies with apes, not only in the United States, but throughout the world [13]. Not as well recognized was Dr. Yerkes' strong support and advocacy of medical research with chimpanzees and other primates. For example, in 1916 in the journal *Science*, Yerkes wrote: 'I am wholly convinced that the various medical sciences and medical practices have vastly more to gain than has yet been achieved, or than any considerable number of medical experts imagine, from the persistent and ingenious use of the monkeys and the anthropoid apes in experimental inquiry' [3].

Yerkes was a visionary with regard to the important uses that chimpanzees and the other primates would eventually be put to in the interests of medicine. It also should be noted that Sir Solly Zuckerman, renowned anatomist at the University of London during the early part of this century, also predicted the value of primates to medical research. In his 1933 book, *The Social Life of Monkeys and Apes*, he wrote: 'Monkeys and apes are chosen as experimental subjects largely because they are the only animals suitable for the investigation of certain diseases, for example, poliomyelitis, or for the analysis of physiological mechanisms such as the menstrual cycle' [14].

Before Dr. Yerkes established his primate research institute in 1930, a major focus of studies with primates was anatomy, particularly comparative anatomical and taxonomic studies using cadavers and skeletons of chimpanzees and primates [4, 15]. While such studies revealed the similarities of humans and chimpanzees in structure, Yerkes believed that similarities also existed in biological function as well and that it was these similarities, once documented, that would benefit the medical sciences [3, 11]. The Laboratories of Primate Biology began with 33 chimpanzees [1]. Today, the Yerkes Center has over 200 chimpanzees as a result of breeding programs, as well as 1,800 other primates of fourteen species at the Yerkes Main Station and Field Station. Yerkes paid great attention to the quality of life of chimpanzees at the research institute and established detailed regulations for laboratory conditions and the care of the animals. He had visited Africa to determine how chimpanzees lived in the wild and developed ideas about how the captive situation could correspond, insofar as possible, with the conditions of nature [1]. Yerkes also sponsored the first field studies of chimpanzees [16], as well as gorillas [17], in large part to gather information that could help in the care and breeding of these animals in captivity [4]. Indeed, the Orange Park institute generated basic information about primate husbandry which improved the care of captive chimpanzees and other primates everywhere, particularly in zoos, whose conditions, Dr. Yerkes wrote in 1924, 'are supposed to be on the whole very good for the life of captive animals'. He noted, however, that the life span of the orangutan and chimpanzee in zoos was then 'measured in months instead of years as of course it should be' [10].

The primary focus of the early years of chimpanzee research at the institute was behavior, particularly learning and intelligence. There researchers developed a battery of tests to study systematically the chimpanzee's mentality, including visual and auditory acuity, visual contrast, problem solving, food sharing and reasoning abilities of many types [1, 18–28]. One of these behavioral tests showed that a chimpanzee could understand the relationship between a key and lock and use the key to unlock a container [29]. Over 50 years later, Yerkes scientists showed that a young chimpanzee not only could understand this relationship but could also use a computer-operated keyboard of symbols which could be pressed to request and obtain from another chimpanzee the appropriate key or tool to unlock a container of food [30]. The goal of the early studies was, in part, to determine the place of the chimpanzee in the phylogeny of other primates and humans [31] and to determine the extent of its behavioral abilities compared to those of humans, particularly children [1].

In 1943, 13 years after his institute had opened, Yerkes wrote that the '... results of experimental inquiry justify, I believe, the working hypothesis that processes other than those of (behavioral) reinforcement and inhibition, function in chimpanzee learning.... I suspect that they presently will be identified as antecedents of human symbolic processes. Thus, we leave the subject at a most exciting state of development, when discoveries of moment seem imminent' [1]. This prediction was borne out in subsequent years. In the chimpanzee the symbolic processes of abstraction, generalization, complex reasoning and problem solving far surpass those of the monkeys and in certain regards approach those of the young, developing human. Thirty years later, at the present Yerkes Center, sophisticated linguistic studies with chimpanzees demonstrated that chimpanzees can develop a vocabulary of word symbols in a way that goes far beyond classical Pavlovian and operant conditioning and that chimpanzees can achieve abstract representation, generalization and processes verging on syntax and grammar [32–34]. These same methods developed with our chimpanzees are now used with retarded children to instruct them in language [35, 36].

In genetics, Yerkes and his colleagues also conducted chromosomal studies. The first report that the number of chimpanzee chromosomes was 48 was published in 1940 authored by Yerkes with C.H. Yeager and T.S. Painter [37]. Although there is a striking similarity between the karyotypes of humans and chimpanzees, we now know, of course, that humans have one pair of chromosomes less than do chimpanzees.

Chimpanzees and other primates were rarely studied in medical research in the early years of this century. One of the few scientists of this early era who did conduct medical research with primates was I.O. Mechnikov, a Russian who lived and worked in France. As Yerkes noted in 1943: 'Even before my ideas (for a primate research institute) became (a reality), the eminent Russian medical investigator Mechnikov ... supplied with monkeys and apes from the African countries of his adopted country (France), used them for important studies of human disease. Before 1910, he (Mechnikov) had become convinced of their high value as experimental subjects ...' [1].

In a speech given in 1968, Dr. Boris Lapin, Director of the Soviet Primate Center, stated that Mechnikov 'managed to reproduce models of human diseases which his contemporaries had failed to reproduce in animals including syphilis, enteric fever, diphtheria, etc'. During a 29-year period beginning in 1886, Mechnikov carried out medical studies involving 500 apes and monkeys [38]. It should also be noted that important early behavioral studies with chimpanzees were conducted by a Russian scientist, the psychologist N.S. Ladygina-Kohts, who investigated the sensory and perceptual processes of a young chimpanzee and a human child [39–41]. These studies attracted the attention of Russian biologists who in 1927 established the Sukhumi Primate Center on the Black Sea, which Dr. Lapin now heads. Dr. Yerkes corresponded with both Russian scientists and acknowledged that they greatly influenced his ideas for his own institute. He wrote: 'Knowledge of Mechnikov's observations and conclusions influenced my planning and in particular supported my belief that the chimpanzee should prove uniquely valuable to biological investigators' [1]. Mechnikov's studies are also believed to have influenced Germany to establish an important German facility for chimpanzee studies at Tenerife in the Canary Islands in 1912, for research on physiology and psychology. Dr. Wolfgang Kohler was the chief investigator, spending 5 years there before the station was abandoned after World War I [42]. Yerkes wrote that if the Tenerife station had continued with the expansion of the chimpanzee research facility and program as planned, his own efforts to develop a primate research institute would have been 'inhibited'. The findings from the Tenerife studies of Kohler, Wilhelm Waldeyer and Max Rothmann strengthened Yerkes' conviction that 'anthropoid research is a logical and also an entirely practicable, shortcut to human biology', as he wrote [1].

Studies in reproductive biology and sexual behavior were also a major focus of the early years of research by Yerkes and his colleagues at Orange Park [43–50]. Yerkes and his colleagues observed the importance and frequency of sexual behavior in chimpanzees, and consequently studies were soon directed at understanding the biology and maturation of the reproductive system, the physiology and endocrinology of reproduction, as well as the sexual behavior of chimpanzees. Yerkes was a colleague and a great admirer of Dr. Alfred Kinsey and both were interested in the application of these studies to understanding human sexuality and reproduction [Yerkes Blanshard R., pers. commun.].

From the beginning of the Orange Park laboratory, one of Dr. Yerkes' goals was to establish a self-sustaining colony of chimpanzees so that the population in captivity would be sufficient to meet the needs of both behavioral and biomedical research [1]. Today, in the USA, we have a National Chimpanzee Breeding and Research Program, directed at establishing a self-sustaining population of chimpanzees for research, particularly in infectious diseases, with emphasis on AIDS [51, 52]. As a result, the USA has the 'strongest and most successful captive chimpanzee management program in place in the history of chimpanzee management and care', according to Dr. Michael Keeling, a participant in this symposium [53]. Many of the early studies on chimpanzee reproduction occurred during the 1930s and early 1940s. From the late 1940s to the early 1960s, there were few studies on chimpanzee reproduction reported in the literature. But reproductive studies of chimpanzees were revived in the late 1960s because of the increasing concern for preserving the species, for both its own sake and for the needs of research [54].

In 1941, Robert Yerkes retired as Director of the Yale Laboratories in Orange Park, which Yale University then renamed the Yerkes Laboratories in his honor. He had written 56 scientific publications on chimpanzees alone, totaling over 2,300 pages and providing his colleagues and successive generations of primatologists and other scientists with invaluable information on this species [12].

Neurological studies became an increasing focus of the Yerkes laboratories in the 1940s reflecting the interests of its brilliant new director, Dr. Karl Lashley, a neuropsychologist. Neurological studies, however, were not completely new to the Orange Park laboratories. There, in 1933, Drs. Carl Jacobsen of the University of Chicago and John Fulton of Yale University began neurological research with chimpanzees [55] that influenced the Portuguese neurologist and neurosurgeon Egas Moniz to use so-called psychosurgery in the treatment of mental disorders. In 1949, Dr. Moniz received the Nobel Prize for his work. In 1955, Dr. Lashley told *Scientific American* magazine that '... if the influence on the development of "psychosurgery" ascribed to Jacobsen's experiment is correct, then this single study has been worth more, in terms of the usual cost and returns of psychiatric research, than the entire investment in the construction and maintenance of the Yerkes laboratories' [56]. In these early years, scientists regarded the chimpanzee brain as a microcosm of the human brain and as a tool for understanding the behavioral similarities between humans and primates. Although only about one fourth the size of the human brain, the chimpanzee brain has virtually all of the features of the human brain and shares with humans that unique asymmetry of the left and right temporal lobes (planum temporale) that permits the development of language in humans and is absent in the lower primates. This particular feature likely explains the ability of chimpanzees to learn and use symbols in a linguistic fashion [57, 58].

During the early 1930s, studies provided valuable gross descriptions of ape brain and brain indices. Detailed maps and descriptions of the configuration of the cerebral hemispheres of chimpanzees were published in 1936 [57–59]. Studies of human, chimpanzee and monkey brains provided important comparative data about the organization of the cortex and brain in general. Beginning in the late 1930s, the sensory cortex of the chimpanzee was among those functional maps of nonhuman primate cortices that were constructed, showing similarities and differences with the human cortex. During the 1940s, scientists continued to produce functional maps of nonhuman primate cortices [59–63]. During the 1960s, visual structures and organization within the brain became a major topic of primate mapping studies [64], and a resurgence of interest in comparative neurology studies occurred. Maps of the sensory, motor and associative cortices of primates, including apes, were generated in substantial numbers [59]. Notwithstanding the many research findings in primate neuroanatomy produced in the past 50 years, they still have not enabled physical anthropologists to provide precise or definitive theories regarding specific qualitative changes in the brain that occurred during primate and human evolution [59]. In addition to renewing interest in comparing the brain mechanisms that underlie primate and human behavior, Lashley helped strengthen the ties between primatologists in the USA and Europe by stressing the importance of ethological studies which were then the major focus of European primate work and by tying together the American emphasis on learning and memory with Europe's focus on genetics, evolution and ethology [12].

Lashley's successor as Director in 1955 was Dr. Henry W. Nissen, a psychologist whose rapport with chimpanzees was described by Yerkes as 'second only to Tarzan's' [1]. Nissen is well known not only for his developmental studies of chimpanzees which began in the 1930s but also for his field studies of chimpanzees in Africa. His African studies, begun in 1928, were the first scientific studies of chimpanzees in the wild – conducted in collaboration with the Pasteur Institute in Kindia, French Guinea, Africa [15].

From the 1950s, and particularly in the 1960s, there was a surge in the use of chimpanzees in basic and applied biomedical studies in infectious diseases and immunology. This emphasis increased exponentially at the Yerkes Center and elsewhere as each decade passed, to the point that medical studies today tend to dominate the scientific literature on chimpanzees. Also during the 1950s, chimpanzees, including animals from the Yerkes Laboratories in Orange Park [56], were studied to advance our understanding of poliomyelitis and develop a vaccine. Alfred Sabin, developer of one of the poliomyelitis vaccines, wrote that the 'important phenomenon of intestinal resistance to reinfection' was revealed in studies with chimpanzees and cynomolgus monkeys that had been infected experimentally with poliomyelitis by oral administration of the viral agent. This finding was one of the factors that promoted the search for nonvirulent strains to use in live poliovirus vaccines [65]. During a two-and-a-half year period, beginning in 1954, Sabin reported that he used 150 chimpanzees, about 9,000 monkeys and 133 human volunteers in his poliomyelitis studies [65–67]. The poliomyelitis vaccine was one example of clear benefit of using chimpanzees in the study of an infectious disease. Many years of studies with chimpanzees, beginning in the 1940s, were devoted to understanding infectious hepatitis, leading to the development of the hepatitis B vaccine, testing of the effectiveness and safety of experimental vaccines involved the extensive use of chimpanzees [68–71].

In the 1960s, partially by the use of chimpanzees, a slow virus was identified as the infectious agent for the fatal central nervous system disease Kuru of the Fore people of New Guinea. Kuru was the first slow virus disease demonstrated in humans [72, 73]. For his contributions to this and related work, Dr. Carleton Gajdusek received the Nobel Prize in Medicine and Physiology. As we all know, today chimpanzees are extremely important in studies of the viral hepatitis diseases, AIDS, malaria and other parasitological infections. I need say no more about this aspect of chimpanzee research since it is well covered in other papers of this symposium.

In the 1960s, organ transplantation studies began using chimpanzees and other primates. Chimpanzees were studied in the early 1960s as sources of organs for transplant surgeries not only because of immunological similarities between humans and chimpanzees, but also because of similarities in size, hemoglobin structure, serum proteins and renal function [74, 75]. However, xenograft studies and attempts were made before information on the nature and genetics of chimpanzee histocompatibility antigens, in the early 1960s, were being defined in humans [75]. Chimpanzee kidneys were used experimentally as transplants for humans. The experimental use of chimpanzee kidneys was 9 months, but unfortunately most such renal transplant attempts resulted in rapid and irreversible rejection [76]. However, the possibility that chimpanzee kidneys would be widely used in human transplantations seemed so strong in 1965 that Dr. William Conway, Director of the New York Zoological Society, expressed concern about the impact of this use on the wild population of chimpanzees [77]. Although Yerkes and his colleagues at the Orange Park Institute had demonstrated that chimpanzees could reproduce in captivity and develop in a healthy fashion [1], it required a significant period of time before scientists and administrators realized that the wild population did not have to be the sole source of these animals for biomedical research.

Chimpanzees were also regarded at one time as a potential source of hearts for human transplant surgeries. In 1964, a chimpanzee heart was first used for a human transplant operation [78, 79]. It should be mentioned that the patients in these early experimental transplant surgeries had end-stage disease and most of these xenograft studies occurred before our understanding of the importance of histocompatibility antigens in transplants. In the 1960s, these antigens were in the process of being defined in humans and exact information on chimpanzee genetics was still to be obtained, as indeed it eventually was [78, 79]. More recently, consideration has again been given to the use of chimpanzee organs for human transplants. It is possible, however, that social and ethical concerns and the lack of a ready supply of chimpanzees for use as organ donors will prevent a sustained attempt to use these animals for this purpose [80].

Chimpanzees also have been important to anthropological studies. In 1930, the same year that Yerkes established his primate research institute, the major focus of anthropologists who were studying primate phylogeny was the place of humans in the primate order. It was through comparative anatomy using cadavers and skeletons of primates [81], including the chimpanzee, that questions of phylogeny were initially investigated [30]. The fossil record suggests that African apes – chimpanzees and gorillas – are useful in the understanding of primate and human evolution. Both Huxley and Darwin used comparisons between humans and living apes to create descriptions of how the early hominoids appeared and behaved. It was not until the 1950s and the 1960s that *fossils* of primates and hominoids became of major importance, to a large extent replacing comparative anatomy as the primary method for these evolutionary studies [30]. Today the use of fossils has been complemented and succeeded in primate and human evolutionary studies (but not totally replaced) by studies using new developments in molecular anthropology and molecular phylogenetics, such as immunological comparisons using electrophoresis (introduced in the 1960s), DNA hybridization, protein sequencing and nucleic acid technology [81–83].

Throughout the history of primatology, anthropologists and other scientists have looked to the living primates for clues to how our primitive ancestors behaved. It has been theorized that certain behaviors such as the eating of meat propelled other behavioral changes, such as tool use, which were milestones in our evolution. These anthropologically oriented studies of primates continue today [84–90].

The study of chimpanzees in behavioral and medical research has been a factor influencing the emergence of these animals in science. However, it is not the only factor. Supply, cost of obtaining and caring for these animals and the ability of scientists to conduct experimental manipulations without stressing the animal unduly while protecting the human staff from injury, have also affected the use of these animals [54]. These factors were recognized by Yerkes, who in 1916 blamed 'technical difficulties and costliness of research' for the lack of information about the structure and development, physiological processes, diseases and pathological anatomy, heredity, life history and behavior of primates. 'Most investigators', he wrote, 'are either impelled or compelled by circumstances to work on easily available and readily manageable organisms. Many of the primates fail to meet these requirements for they are relatively difficult and expensive to obtain by importation or breeding and to keep in normal condition' [3]. The development of improved anesthesia, particularly the arrival of ketamine in the 1970s, enabled increased studies using chimpanzees in medical research in which blood and other tissues routinely must be obtained. In fact, improved, short-acting, subcutaneous and intramuscular anesthesia revolutionized the way that chimpanzees could be studied and opened doors to investigations that were previously not deemed possible [54].

In closing, I return to my earlier statement that the Yerkes Primate Center, in many respects, reflects historically the emergence and pattern of use of chimpanzees in research. In the 1930s and 1940s, behavioral studies dominated, with reproductive studies primarily focusing on the behavioral aspects as well as the biology and physiology of reproduction. These were followed by increasingly complete neurological investigations and eventually other more applied medical studies with the emphasis on infectious diseases which have increased rapidly during the subsequent decades. Today, Yerkes' dreams and plans for scientific institutes, focused on both behavioral and biomedical research and utilizing a self-sufficient population of animals, have been realized and I am certain that, if Yerkes were with us today, he would be immensely pleased by this opening of the Hans Popper Primate Center – a further extension of his vision and dream beginning in 1900 and extending through generations of scientists to 1992.

Acknowledgements

We gratefully acknowledge the assistance of Nellie Johns and Margaret Milne in the preparation of this manuscript. The writing of this manuscript was supported in part by NIH grant RR-00165 to the Yerkes Regional Primate Research Center, which is fully accredited by the American Association for the Accreditation of Laboratory Animal Care.

References

1 Yerkes RM: Chimpanzees: A Laboratory Colony. New Haven, Yale University Press, 1943.
2 Carmichael L: The Past, Present and Future of Scientific Primatology; in Carpenter CR (ed): Second International Congress of Primatology. Basel, Karger, 1969, vol 1, pp 1–10.
3 Yerkes RM: Provision for the study of monkeys and apes. Science 1916;43:231–234.
4 Ribnick R: A short history of primate field studies: Old World monkeys and apes; in Spence F (ed): A History of American Physical Anthropology 1930–1970. New York, Academic Press, 1982, pp 49–73.
5 Hooten EA: Man's Poor Relations. New York, Doubleday, 1942.
6 Kroeber AL: Sub-human culture beginnings. Q Rev Biol 1928;3:325–342.
7 Haggerty ME: Preliminary experiments on anthropoid apes. Psychol Bull 1910;7:49.
8 Miller OT: Mr. Crowley, the chimpanzee. Cosmopolitan 1887;4:297–304.
9 Schlater PS: The bald headed chimpanzee. Nature 1889;39:254.
10 Yerkes RM: Almost Human. New York, Century, 1925.
11 Yerkes RM, Morgulis S: The method of Pavlov in animal psychology. Psychol Bull 1909;6:257–273.
12 Mitchell G: In defense of primate psychology. Am Biol Teachers 1988;50:86–90.
13 Dewsbury DA: Comparative Psychology in the Twentieth Century. Shroudsburg, Hutchinson Ross, 1984.
14 Zuckerman S: The Social Life of Monkeys and Apes. London, Routledge and Kegan Paul, 1932.
15 Schultz AH: Changing Views on the Nature and Interrelations of the Higher Primates. Yerkes Newsletter. Atlanta, Yerkes Regional Primate Research Center, Emory University, 1966, vol 3, pp 15–29.
16 Nissen HW: A field study of the chimpanzee, observations of chimpanzee behavior and environment in western French Guinea. Comp Psychol Monogr 1931;8:1–127.
17 Bingham HC: Gorillas in a Native Habitat. Publ No 426. Washington, Carnegie Institution of Washington, 1932, pp 1–66.
18 Jacobsen CF, Jacobsen MM, Yoshioka JG: Development of an infant chimpanzee during her first year. Comp Psychol Monogr 1932;9:1–94.
19 Kellogg WN, Kellogg LA: The Ape and the Child: A Study of Environmental Influence upon Early Behavior. New York, Whittlesey House, 1933.

20 Yerkes RM: Modes of behavioral adaptation in chimpanzee to multiple choice problems. Comp Psychol Monogr 1934;10:1–108.
21 Nissen H, Crawford M: A preliminary study of food sharing behavior in young chimpanzees. J Comp Psychol 1936;22:383–419.
22 Crawford MP: The cooperative solving of problems by young chimpanzees. Comp Psychol Monogr 1937;14:1–88.
23 Spence KW: Analysis of the formation of visual discrimination habits in chimpanzees. J Comp Psychol 1937;23:77–100.
24 Crawford MP: The cooperative solving of problems by young chimpanzees. Comp Psychol Monogr 1937;14:1–88.
25 Yerkes RM: Primate cooperation and intelligence. Am J Psychol 1937;50:254–270.
26 Yerkes RM, Nissen HW: Pre-linguistic sign behavior in chimpanzee. Science 1939;89:585–587.
27 Spence K: The solution of multiple choice problems by chimpanzees. Comp Psychol Monogr 1939;15:1–54.
28 Cowles JT: Food tokens as incentives for learning by chimpanzees. Comp Psychol Monogr 1937; 14:1–96.
29 Yerkes RM, Yerkes AW: The Great Apes: A Study in Anthropoid Life. New Haven, Yale University Press, 1929.
30 Savage-Rumbaugh ES, Rumbaugh DM, Boysen S: Linguistically mediated tool use and exchange by chimpanzees (*Pan troglodytes*). Behav Brain Sci 1978;4:539–554.
31 Fleagle JG, Jungers WL: Fifty years of higher primate phylogeny; in Spence F (ed): A History of American Physical Anthropology, 1930–1980. New York, Academic Press, 1982, pp 187–230.
32 Savage-Rumbaugh ES: Ape Language: From Conditioned Response to Symbol. New York, Columbia University Press, 1986.
33 Savage Rumbaugh ES, Rumbaugh DM, Boysen S: Symbolic communication between two chimpanzees (*Pan troglodytes*). Science 1978;201:641–644.
34 Rumbaugh DM (ed): Language learning by a Chimpanzee: The Lana Project, New York, Academic Press, 1977.
35 Romski MA, White RA, Millen CE, Rumbaugh DM: Effects of computer-keyboard teaching on the symbolic communication of severely retarded persons: Five case studies. Psychol Rec 1984;34:39–54.
36 Romski MA: Two decades of language research with great apes. Am Speech Lang Hear Assoc 1989, pp 81–82.
37 Yeager CH, Painter TS, Yerkes RM: The chromosomes of the chimpanzee. Science 1940;91:74–75.
38 Lapin BA: Development of Experimental Primatology in Russia and in the USSR. Yerkes Newsletter. Atlanta, Yerkes Regional Primate Research Center, Emory University, 1968, vol 5, pp 19–22.
39 Ladygina-Kohts NN: Untersuchungen über die Erkenntnisfähigkeiten des Schimpansen (in Russian with German translation of summary). Moscow, Museum Darwinianum, 1923.
40 Ladygina-Kohts NN: The Constructive and Tool Making Activation of Higher Primates (Chimpanzees). Moska, Ixdutelstvo Akademii Nauk, 1959.
41 Yerkes RM, Petrunkevich A: Studies of chimpanzee vision by Ladygina-Kohts. J Comp Psychol 1925;5:99–108.
42 Kohler W: The Mentality of Apes. London, Routledge and Kegan Paul, 1925.
43 Bingham HC: Sex development in apes. Comp Psychol Monogr 1928;5:1–165.
44 Tinklepaugh OL: Sex cycles and other cyclic phenomena in a chimpanzee during adolescence, maturity and pregnancy. J Morphol 1933;54:521–547.
45 Yerkes RM: Multiple births in anthropoid apes. Science 1934;79:430–431.
46 Yerkes RM: A second generation captive born chimpanzee. Science 1935;81:542–543.
47 Tomilin MI, Yerkes RM: Chimpanzee twins: Behavioral relations and development. J Genet Psychol 1935;46:239–263.
48 Elder JH: The time of ovulation in the chimpanzee. Yale J Biol Med 1938;10:347–364.
49 Yerkes RM: Sexual behavior in the chimpanzee. Hum Biol 1939;11:78–111.
50 Nissen HW, Yerkes RM: Reproduction in the chimpanzee: Report of 49 births. Anat Rec 1943;86: 567–578.
51 Johnsen DO: The need for using chimpanzees in research. Lab Animal 1987;16:19–23.
52 National Center for Research Resources: Progress Report on NIH Chimpanzee Breeding and Research Program. Rockville, National Institutes of Health, 1990.
53 Keeling ME: Letter to Chief, Office of Scientific Authority. Washington, US Fish and Wildlife Service, 1992.

54 Keeling ME, Roberts JR: Breeding and reproduction of chimpanzees; in Bourne GH (ed): The Chimpanzee. Basel, Karger, Baltimore, University Press, 1972, vol 5, pp 127–152.
55 Fulton JF, Jacobsen CF: The functions of the frontal lobes, a comparative study in monkeys, chimpanzees and man. Adv Mod Biol Moscow 1935;4:113–123.
56 Gray GW: The Yerkes laboratories. Sci Am 1955, pp 2–12.
57 Le Gros Clarck WE, Coper DM, Zuckerman SO: The endocranial cast of the chimpanzee. J R Anthropol Inst Great Britain Ireland 1936;66:249–268.
58 Walker E, Fulton JF: The external configuration of the cerebral hemispheres of the chimpanzee. J Anat Lond 1936;71:105–116.
59 Falk D: Primate neuroanatomy: An evolutionary perspective; in Spence F (ed): A History of American Physical Anthropology, 1930–1980. New York, Academic Press, 1982, pp 75–103.
60 Dusser de Barenne JG, Garol JG, McCulloch WS: The sensory cortex of the chimpanzee. Proc Soc Exp Biol Med 1939;42:27–29.
61 Bailey P, Dusser de Barenne JG, Garol HW, McCulloch WS: The sensory cortex of the chimpanzee. J Neurophysiol 1940;3:469–485.
62 Bailey P, Von Bonin G, Garol H, McCulloch WS: Long association fibers in cerebral hemispheres of monkey and chimpanzee. J Neurophysiol 1943;6:129–134.
63 Woolsey CN, Marshall WH, Bard P: Note on organization of tactile sensory area of cerebral cortex of chimpanzee. J Neurophysiol 1943;6:287–291.
64 Woolsey CN, Marshall WH, Bard P: Organization of pre- and postcentral areas in chimpanzee and gibbon. Trans Am Neurol Assoc 1960;85:144–146.
65 Sabin AB: Oral poliovirus vaccine: History of its development and use and current challenge to eliminate poliomyelitis from the world. J Infect Dis 1985;151:420–436.
66 Sabin AB: Immunization of chimpanzees and human beings with avirulent strains of poliomyelitis virus. Ann NY Acad Sci 1955;61:1050–1056.
67 Sabin AB: Behavior of chimpanzee-avirulent poliomyelitis viruses in experimentally infected human volunteers. Am J Med Sci 1955;230:1–8.
68 Maynard JE, Berquist KR, Krushak DH, Purcell RH: Experimental infection of chimpanzees with the virus of hepatitis B. Nature 1972;237:514–515.
69 Prince AM: Use of chimpanzee as a model for the study of hepatitis B virus infection; in Goldsmith EI, Moor-Jankowski J (eds): Medical Primatology. Basel, Karger, 1972, p 97.
70 Barker LF, Chisari FV, McGrath PP, Dalgard DW, Kirschstein RL, Almeida JD, Edgington TS, Sharp DG, Peterson MR: Transmission of type B viral hepatitis to chimpanzees. J Infect Dis 1973; 127:648–654.
71 Purcell RH: Primates and hepatitis research, in Erwin J, London JC (eds): Chimpanzee Conservation and Public Health, Environments for the Future. Rockville, Diagnon/Bioqual, 1992, pp 15–20.
72 Leader RW: The kinship of animal and human diseases. Sci Am 1967;216:110–116.
73 Gajdusek DC, Gibbs CJ, Alpers M: Experimental transmission of a kuru-like syndrome to chimpanzees. Nature 1966;209:794.
74 Reemtsmi K: Heterotransplantation. Yerkes Newsletter. Atlanta, Yerkes Regional Primate Research Center, Emory University, 1969, vol 6, pp 18–21.
75 Metzger RS, Seigler HF: Transplantation biology of the chimpanzee; in Bourne GH (ed): The Chimpanzee. Basel, Karger, 1970, vol 3, pp 20–25.
76 Reemtsmi K: Xenotransplantation: A personal history; in Hardy MA (ed): Xenograft. Elsevier, 1989, vol 25, pp 7–16.
77 Conway WG: The availability of long-term supply of primates for medical research: a report on the conference held in New York. Int Zoo Year, Zool Soc Lond, 1966;6:284–288.
78 Hardy JD, Chavez CM, Kurras JD, et al: Heart transplantation in man. JAMA 1964;188:114.
79 Lansman SL, Ergin MA, Griepp RB: The history of heart and heart-lung transplantation. Cardiovasc Clin 1990;20:3–19.
80 Fletcher JC, Robertson JA: Primates and ancephalics as sources for pediatric transplants. Fetal Ther 1986;1:150–164.
81 Sibley CG, Ahlquist JE: The phylogeny of the hominoid primates, as indicated by DNA-DNA hybridization. J Mol Evol 1984;20:2–15.
82 Sarich VM, Wilson AC: Immunological time scale for hominid evolution. Science 1967;158:1200–1203.
83 Marks J: What's old and new in molecular phylogenics. Am J Phys Anthropol 1991;85:207–219.
84 Boesch C, Boesch H: Hunting behavior of wild chimpanzees in the Tai National Park. Am J Phys Anthropol 1989;78:547–573.

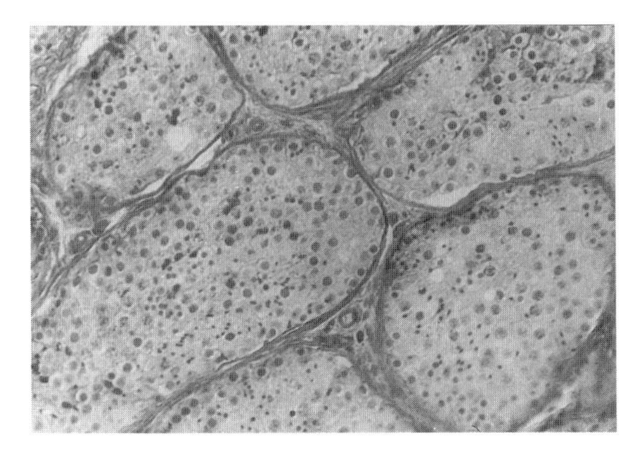

Fig. 3. Testicular histology of the chimpanzee. Note the apparent absence of spermatozoa within any clearly demarcated lumen within the seminiferous tubule.

that seen in a number of other primate species. There is some evidence, however, that is indicative of a seasonal change in semen motility parameters which may be associated with birth rate (fig. 4).

Techniques have been developed for the routine collection of semen from the chimpanzee male [41, 42] utilizing recovery of semen samples by a variety of methods. These include recovery after deposition within or near the female tract [35], by collection at the time of masturbation [39, 43, 44], by collection into an artificial vagina [45] or collection at the time of electrical stimulation, either of the penis of a conscious male [46, 47] or of the pelvic organs of an anesthetized male [30, 48]. The quality of the collected semen varies according to the collection method used [49, 50], and between individual males.

Methods are now available for sophisticated analysis of sperm viability, motility, fertilizing capacity and membrane composition [51–55]. Computer-assisted motion analysis (CAMA) of sperm motion characteristics is practical using videotape recordings as they are based upon a digital image, derived from a charge-coupled device with functional dimensions measured in pixel units. Earlier reports quantifying sperm motion were based on painstaking measurement and analysis of individual frames of a ciné or videotape record, or used computer-assisted plotting of sperm motion on a frame-by-frame basis [56, 57]. Current CAMA makes it practical to analyze a sufficient number of samples/individual sperm to be of use in the determination of motility parameters that may be correlated with potential fertilizing capacity [53, 58, 59]. There are several commercially available CAMA systems, and the data presented here are derived from Motion Analysis Systems (Santa Rosa, Calif., USA) equipment

85 Ghiglieri MP: Sociobiology of the great apes and the hominid ancestor. J Hum Evol 1987;16:319–357.
86 Wrangham RW: The significance of African apes for reconstructing human social evolution; in Kinzey WG (ed): The Evolution of Human Behavior: Primate Models. New York, SUNY Press, 1986, pp 51–71.
87 Kortlandt A: Use of stone tools by wild living chimpanzees and the earliest hominids. J Hum Evol 1986;15:77–132.
88 McGrew WC: Evolutionary implication of sex differences in chimpanzee predation and tool use; in Hamburg DA, McCown ER, (eds): The Great Apes. Menlo Park, Benjamin/Cummings, 1979, pp 440–463.
89 Bauer HR: Chimpanzee society and social dominance in evolutionary perspective; in Strayer FF, Freedman D (eds): Dominance Relations: Ethological Perspectives in Human Conflict. New York, Garland, 1980, pp 97–119.
90 De Waal F: Chimpanzee Politics. New York, Harper & Row, 1982.

Frederick A. King, PhD, Yerkes Regional Primate Research Center, Emory University, Atlanta, GA 30322 (USA)

Eder G, Kaiser E, King FA (eds): The Role of the Chimpanzee in Research.
Symp, Vienna 1992. Basel, Karger, 1994, pp 121–133

Pathology of the Chimpanzee in Research Facilities[1]

Harold M. McClure, Daniel C. Anderson, Sherry A. Klumpp

Yerkes Regional Primate Research Center, Emory University, Atlanta, Ga., USA

Introduction

The chimpanzee has long been an important laboratory animal for use in various aspects of biomedical and behavioral research and will continue to be an important model system for evaluation of various human disease entities to which only humans and chimpanzees are susceptible. Our ability to continue using the chimpanzee in biomedical and behavioral research will be dependent upon the continued successful maintenance of domestic breeding programs and the control of infectious diseases and other disease-causing agents in chimpanzees maintained in domestic breeding colonies as well as animals maintained in research facilities. Appropriate management practices, preventive medicine programs and clinicopathologic diagnostic support will have a major impact on the control of spontaneous diseases in this unique laboratory animal species.

In this paper, we briefly review the literature related to the diseases and pathology of chimpanzees and summarize the clinicopathologic data derived from chimpanzees at the Yerkes Regional Primate Research Center for the period from 1967 to 1991. During this 25-year period, 186 chimpanzees were necropsied and hematologic, blood chemistry and parasitologic evaluations were conducted routinely on chimpanzees with clinical disease problems. Biopsies, cerebrospinal fluid examinations and other specialized tests were conducted as needed.

[1] This work was supported by NIH grant RR00165 from the National Center for Research Resources to the Yerkes Regional Primate Research Center. The Yerkes Center is fully accredited by the American Association for Accreditation of Laboratory Animal Care.

Colony and Mortality Statistics

During the 25-year period covered by this paper and due primarily to a very successful breeding program, the size of the chimpanzee colony at the Yerkes Center increased from 78 animals in 1967 to 206 animals in 1991. The number of chimpanzees in the colony at various time points during this 25-year period was as follows: 1967, 78; 1969, 77; 1974, 101; 1979, 126; 1984, 141; 1989, 178; 1990, 202; 1991, 206.

The 186 chimpanzees necropsied during this time period were categorized, in general, as follows:

General category	Chimpanzees	
	n	%
Deaths associated with clinical problems	54	29.0
Abortuses and stillbirths	45	24.2
Neonatal deaths	42	22.6
Deaths associated with experimental procedures	40	21.5
Deaths due to accidents or trauma	5	2.7

As evident from the above numbers, 46.8% of the chimpanzees necropsied at the Yerkes Center during the past 25 years were either abortuses, stillbirths or neonatal deaths (87 of 186 necropsies). During this period, 278 chimpanzees were born at the Yerkes Center; this included 45 (16%) stillbirths or abortuses and 233 (84%) livebirths. 42 (18%) of the liveborn animals died during the neonatal period. The age of chimpanzees that died during the neonatal period is summarized as follows:

Age at death	Animals		Accumulative percent
	n	%	
Day of birth	26	61.9	61.9
Days 1–7	6	14.3	76.2
Days 8–14	6	14.3	90.5
Days 15–21	4	9.5	100

The primary cause of death in chimpanzees that died during the neonatal period is summarized as follows:

Cause of death	Animals	
	n	%
Trauma	19	45.2
Prematurity/anoxia	12	28.6
Dehydration/inanition	3	7.1
Septicemia	2	4.8
Pneumonia	2	4.8
Enterocolitis	4	9.5

Bacterial infections associated with the neonatal deaths included enteropathogenic *Escherichia coli* in 2 animals with enterocolitis, 1 enteropathogenic *E. coli* septicemia [1], 1 *Streptococcus pneumoniae* septicemia, 1 *Campylobacter jejuni* enterocolitis and *E. coli* was isolated from the lungs in 1 infant with pneumonia.

During this 25-year period, 54 chimpanzee deaths that occurred beyond the neonatal period were associated with clinical disease problems. The age range of these 54 animals included 7 less than 1 year of age, 12 from 1 to 10 years old, 18 that were 10–20 years old, 12 that were over 20 years of age and 5 adults of unknown age. The oldest chimpanzee in this series was 56 years of age. The major causes of death in this group of animals included: bacterial infections of the respiratory tract, gastrointestinal system and central nervous system (CNS); parasitic infections of the gastrointestinal tract by organisms such as *Strongyloides, Enterobius* and *Balantidium coli*, and cardiovascular diseases such as myocardial fibrosis, brain hemorrhage and pulmonary thromboembolism.

The remainder of this paper will summarize the clinical disease problems encountered in this necropsy series, based on the etiologic agent, including bacterial, viral, parasitic and mycotic infections and neoplasms and miscellaneous entities.

Bacterial Diseases

The major bacterial diseases encountered during this 25-year period included 4 cases of *Mycobacterium tuberculosis* infection, 7 animals with CNS abscesses, meningitis or ventriculitis (1 *E. coli* and β-*Streptococcus*, 1 *Staphylococcus aureus*, 1 *Haemophilus influenzae* and 4 *S. pneumoniae*); 5 animals with pneumonia (2 *S. aureus*, 1 *S. pneumoniae*, 1 with *Streptococcus viridans, E. coli* and *Proteus* sp. isolated and 1 culture negative); 8 animals with septicemia (2 *S. pneumoniae*, 2 *S. aureus*, 3 *E. coli* and 1 *Yersinia enterocolitica*) and 7 cases of enterocolitis (2 *C. jejuni*, 1 *C. fetus* and *C. coli*, 1 enteropathogenic *E. coli*, 1 *Shigella flexneri*, 1 *Y. enterocolitica* and 1 animal from which only *E. coli* was isolated). Animals with CNS infections or pneumonia were often septicemic.

During the past 25 years only 4 cases of tuberculosis were diagnosed in the Yerkes chimpanzee colony. These cases and the date of occurrence can be summarized as follows:

Date	Origin of chimpanzee	Time in colony months	Organs/tissue affected
11–5–70	other colony	5	disseminated
11–16–70	other colony	6	lung and hilar nodes
11–21–70	other colony	18	lung and hilar nodes
2–4–71	other colony	9	lung and hilar nodes

The 3 latter cases were all secondary to the first case, and all 3 were detected by routine tuberculin tests. The first case occurred in a chimpanzee from another colony, 2 months after it was released from quarantine at the Yerkes Center (5 months after receipt at the Yerkes Center). This chimpanzee was repeatedly tuberculin negative and at the time of death was found to have disseminated, miliary tuberculosis which included lesions in the lungs, kidneys and mucosa of the gastrointestinal tract. *M. tuberculosis* was isolated from all 4 chimpanzee cases of tuberculosis.

Although tuberculosis in nonhuman primates is less frequent than in previous years when large numbers of nonhuman primates were imported from the wild, this disease should always be considered a potential threat to nonhuman primate colonies. The resurgence of tuberculosis in the human population throughout the world will increase the risk for exposure of nonhuman primates. An active tuberculosis surveillance program, in both the animal colony and in personnel who have contact with nonhuman primates, should be maintained to prevent infection in domestic nonhuman primate colonies.

Yersiniosis is being recognized with increasing frequency in both humans and nonhuman primates and may be caused by infection with either *Y. enterocolitica* or *Y. pseudotuberculosis*. This disease may occasionally be fatal in both humans and nonhuman primates. Numerous clinical and fatal cases of yersiniosis have been diagnosed in outdoor housed monkeys at the Yerkes Center and a number of clinical isolates have been obtained from chimpanzees with diarrheal disease [2, 3]. However, only 1 fatal case of yersiniosis has been diagnosed in a chimpanzee. The latter occurred in a 13-year-old, male, laboratory-born chimpanzee, approximately 1 month after the animal had been transferred from the Main Center to an outdoor housed chimpanzee breeding colony at the Yerkes Field Station. This animal was found comatose in the outdoor compound; he was dehydrated and had bloody diarrhea. A complete blood cell count revealed a hematocrit of 65.3% and a WBC count of 16,600 with 39% segmented neutrophils, 45% band neutrophils, 7% lymphocytes and 6% monocytes. Bacterial culture of a rectal swab yielded *Y. enterocolitica* and *C. fetus* and *Y. enterocolitica* was also isolated from a blood culture. Despite intensive antibiotic treatment and supportive care, the animal's clinical condition progressively deteriorated. He was euthanatized 6 days after clinical signs had first been noted due to multisystem failure characterized by complete anuria, hepatic failure, thrombocytopenia and enteric bleeding. At necropsy, there was a prominent, hemorrhagic, necrotizing colitis and typhlitis and lymphadenopathy of the mesenteric lymph nodes.

The *Yersinia* isolate from this case was biotype 4 and serotype 0:3, was positive for Congo red uptake and calcium dependency and negative for pyrazinamidase activity (determinations done by Dr. C. Krishnan, Laboratory Services Branch, Toronto, Ontario, Canada). This fatal case of yersiniosis in a chimpanzee occurred during an outbreak of yersiniosis in great apes at the Yerkes Center that included 4 orangutans, 1 gibbon, 2 gorillas and 5 chimpanzees. The isolates from all of these clinical cases were the same biotype and serotype as the isolate from the fatal case in the chimpanzee. This was the first cluster of *Y. enterocolitica* infection due to serogroup 0:3 encountered at the Yerkes Center. Only 3 previous isolates from monkeys were serotype 0:3. Since the mid 1980s, *Y. enterocolitica* 0:3 has become the predominant serotype in the USA and evidence indicates that swine may be the major reservoir for this serotype [4]. The source of infection in Yerkes chimpanzees and other great apes was not identified.

Other bacterial infections that may be encountered with appreciable frequency in laboratory housed chimpanzees include *Campylobacter* species, *β-streptococcus, S. aureus, S. pneumoniae, H. influenzae* and *Shigella* species. The most frequently encountered problems associated with infections of this type during this 25-year period in the Yerkes Center chimpanzee colony included CNS or pulmonary infections. Animals with CNS or pulmonary infections, especially with *S. pneumoniae*, are often septicemic [2].

Naturally acquired leprosy has been reported in at least 3 chimpanzees in the USA [5–7] and serologic surveys of chimpanzee colonies have identified additional chimpanzees with evidence of exposure to *Mycobacterium leprae* [6]. A recent survey of the Yerkes chimpanzee colony has revealed that at least 4 chimpanzees show serologic evidence of exposure to *M. leprae* (work done by Dr. Bobby Gormus, Tulane Primate Research Center). Studies are in progress to confirm this possible *M. leprae* exposure. These recent reports suggest that transmission of *M. leprae* may have occurred in domestic chimpanzee colonies [5–7].

Viral Diseases

Although there were no deaths attributed to viral diseases in the Yerkes chimpanzee colony during this 25-year period, episodes of clinical disease due to infection with *Herpesvirus hominis* type 2 or a varicella-like virus were documented [8–10]. Oral focal epithelial hyperplasia has been documented in adult chimpanzees in the colony, and a small number of chimpanzees have been found to be seropositive for hepatitis C or simian T-lymphotropic virus type 1 (STLV-1).

One episode of *H. hominis* type 2 infection involved 2 adult common chimpanzees and 3 adult pygmy chimpanzees and the second outbreak involved 2 juvenile common chimpanzees in a breeding colony. In the first outbreak, lesions were evident on the external genitalia or oral cavity of the affected chimpanzees. Histologically, biopsy specimens from these lesions revealed typical herpetic changes which included necrosis, superficial ulceration, acute inflammatory cell infiltrates, multinucleated syncytial cells and intranuclear inclusions. Herpes-type viruses were readily demonstrated by electron microscopy of these lesions [10]. Similar lesions were noted in the second outbreak and herpes-type viruses were also readily demonstrated by electron microscopy.

One outbreak of a varicella-like disease (chickenpox) was documented in a group of nursery-reared chimpanzees. The affected chimpanzees showed numerous widespread vesicular lesions on the skin and oral mucosa. Herpes-type virus was identified by electron microscopy of vesicular fluid from the skin lesions. The affected chimpanzee infants were essentially clinically normal except for the skin lesions and all made an uneventful recovery [8].

A number of cases of oral focal epithelial hyperplasia have been seen in adult chimpanzees of the Yerkes colony. This disease, caused by a papillomavirus, has been reported in other chimpanzee facilities. Gross lesions are limited to the oral cavity and consist of variable-sized mucosal elevations that are usually 0.1–0.5 cm in diameter. Histologically, the lesions are characterized by mild to severe focal acanthosis and nuclear degeneration of epithelial cells, with ballooning, clumping and margination of chromatin. Intranuclear papovavirus particles, 45–55 μm in diameter, can usually be found by electron microscopy in epithelial cells of the upper stratum spinosum. The lesions are usually self-limiting but may last for months to years. The prevalence of this disease in chimpanzees is probably greater than is recognized due to the benign, inconspicuous nature of the lesions. One report of several cases in a domestic chimpanzee colony suggested that the disease was communicable within this colony. In humans, DNA hybridization studies have shown an association between focal epithelial hyperplasia and human papillomaviruses (HPV) types 13 and 32 and an oral papillomavirus related to human papillomavirus type 13 was recently described in focal epithelial hyperplasia of pygmy chimpanzees [11].

STLV-1, a virus infecting Old World nonhuman primates, is closely related to human T lymphotropic virus type 1, the causative agent of adult T cell leukemia. Although the effects of STLV-1 infection in great apes are unknown, serologic evidence of STLV-1 infection has been reported in a gorilla with lymphoma and in 3 gorillas that died as the result of chronic, progressively debilitating disease [12]. A recent serologic survey of the Yerkes great ape colony revealed serologic evidence of STLV-1 infection in 1 of 21 gorillas, 0 of 30 orangutans, 3 of 13 pygmy chimpanzees and 10 of 191 chimpanzees. The seropositive animals were clinically normal at the time serum was collected and have continued to remain clinically normal. The long-term effects of STLV-1 infection in chimpanzees and other great ape species is unknown at the present time.

Serologic evidence of infection with hepatitis C virus (HCV) has recently been documented in 11 chimpanzees in the Yerkes Center colony. These 11 ELISA-positive animals were further evaluated by recombinant immunoblot assay, with 5 characterized as nonreactive, 2 as indeterminate and 4 as reactive. These infections were apparently related to previous immunology and/or malaria studies which involved inoculation of some chimpanzees with material derived from humans (before serologic tests for HCV were available). One of the HCV-positive chimpanzees has recently died from other causes; liver sections from this animal did not show any histologic evidence of hepatitis. In a recent survey of another chimpanzee colony for evidence of HCV infection, 7 of 139 animals were found to be seropositive, and 10 additional animals had borderline values [13]. Considerable experimental work has been done with HCV in chimpanzees, and 3 patterns of viremia have been demonstrated. These include (1) transient viremia in acute resolving hepatitis, (2) persistent viremia in chronic hepatitis and (3) intermittent viremia in chronic hepatitis [14]. In chimpanzees with chronic HCV infection, the HCV genome can be consistently detected by polymerase chain reaction for up to 10 years after infection [15].

Parasitic Diseases

Parasites encountered in this necropsy series included *Enterobius, Strongyloides, Trichuris, Balantidium coli, Gongylonema, Schistosoma* and *Capillaria hepatica* [8]. Other parasites encountered clinically during this period included *Giardia* and *Entamoeba* spp. These parasites were often incidental findings that did not contribute significantly to clinical disease or death of the animal. However, occasional severe cases of enterobiasis, strongyloidiasis and balantidiasis were encountered. *Pneumocystis carinii* pneumonia was diagnosed in 2 young chimpanzees that were being used in a bovine leukemia virus study. Incidental *Schistosoma mansoni* infection has been reported in a chimpanzee [16], as have severe or fatal cases of amebiasis [17, 18] and enterobiasis [19, 20].

A fatal case of enterobiasis occurred in a 12-year-old male chimpanzee. This animal developed pneumococcal pneumonia and meningitis (*S. pneumoniae* type 18) for which he was successfully treated with antibiotics, fluids and electrolytes. Two months after recovery from the meningitis, the animal developed severe diarrhea and died. Necropsy revealed a severe enterocolitis and peritonitis associated with *Enterobius vermicularis* infection. Histologically, there was extensive mucosal and submucosal necrosis with infiltration of polymorphonuclear leukocytes. Occasional parasites were evident in the mucosa, with extremely large numbers of parasites present in the submucosa. A few parasites were evident on the serosa. The parasites were identified as *E. vermicularis*, based on characteristic prominent lateral alae and a large, posterior esophageal bulb [20].

Although hyperinfection with *Strongyloides* has been a more frequent problem in young orangutans in the Yerkes colony than in chimpanzees [21], a number of cases have been seen in chimpanzees. Affected animals may have diarrhea, with *Strongyloides* larvae identified in fecal samples, or may present with a nasal discharge and signs of a severe pulmonary disease. The latter is associated with extensive hemorrhage into the alveolar spaces in the lungs due to migration of *Strongyloides* larvae through the lung parenchyma. In cases of hyperinfection, numerous *Strongyloides* larvae may be evident in the mucosa and wall of the intestine, as well as in mesenteric lymph nodes, the spleen, kidneys, liver and lung and in potentially any other tissue of the body, including the brain.

B. coli is found with considerable frequency in fecal examinations of chimpanzees and occasional *B. coli* can be found in the superficial mucosa of sections of the colon of chimpanzees that have died from various causes. *B. coli* is, therefore, usually an incidental finding with little evidence for pathogenicity in the chimpanzee. However, in some clinical cases of diarrhea, extremely large numbers of *B. coli* are found in fecal samples suggesting that this organism may be of etiologic significance. In addition, histologic evaluation of ulcerative lesions in the colon will, on occasion, reveal numerous *B. coli* that have invaded deeply into the mucosa and submucosa of areas of ulceration. In such cases, the *B. coli* is often secondary to some other primary pathogen.

P. carinii pneumonia was diagnosed in 2 infant chimpanzees, both of which had an underlying myeloproliferative disorder [22]. In addition to the hematologic abnormalities associated with the myeloproliferative disorder, both of these chimpanzees presented clinically with a pneumonia, severe dyspnea, anorexia and inactivity. Neither animal showed a significant response to therapy; 1 animal died at 34 weeks of age, 6 weeks after the onset of signs of pneumonia and the other animal died at 45 weeks of age, 5 weeks after the onset of pneumonia.

At necropsy, the lungs of 1 of these chimpanzees were diffusely consolidated, dark red and contained scattered subpleural areas of hemorrhage. In the other case, the lungs contained numerous, scattered discrete areas of consolidation. When sectioned, the lungs had a characteristic dry, doughy consistency. Histologically, the lungs showed extensive interstitial pneumonia with a mononuclear cell infiltrate. Alveolar spaces in 1 case were filled with eosinophilic, foamy material; this material was not present in the alveolar spaces of the second case which had been treated with pentamidine isethionate. Gomori methenamine silver stain revealed numerous *P. carinii* organisms in the alveolar spaces of both cases [22, 23].

Mycotic Diseases

Mycotic diseases have been infrequently reported in the chimpanzee. Reported mycotic infections include *Microsporum* and *Trichophyton* infections of the skin [24, 25], piedra [26], rhinophycomycosis [27, 28], coccidioidomycosis [29] and candidiasis [30]. The only mycotic infection observed in this series was 1 case of oroesophageal candidiasis that occurred in a chimpanzee with diabetes.

Neoplasms

Neoplastic diseases have been infrequently reported in chimpanzees, probably due in large part to the small number of chimpanzees, particularly older chimpanzees, that have been subjected to a complete necropsy examination. In 1 series of 268 chimpanzee necropsies over a 15-year period, no neoplasms were observed [31]. In another survey of a chimpanzee colony, the only neoplastic-type condition observed over a 10-year period was 1 case of nasal polyps [32]. The few neoplasms that have been reported in chimpanzees, some of which were poorly documented, include a brain tumor of undetermined type, cutaneous papillomas, odontoma of the maxillary area and nasal polyps [8]. One recent report of a hepatocellular carcinoma in a chimpanzee was believed to be associated with a non-A, non-B hepatitis virus infection [33].

The only neoplasms diagnosed over a 25-year period in the Yerkes chimpanzee colony included an ovarian fibrothecoma in a 48-year-old female, diffuse adenomatosis of the islets of the pancreas of a 47-year-old female, a subcutaneous lipoma in a 29-year-old female, and a hemangioma of the subcutis in a 2-month-old female [34]. Two cases of myeloproliferative disease (erythroleukemia) encountered in 2 young chimpanzees in this series were believed to be related to experimental procedures [22].

Miscellaneous Diseases

A variety of miscellaneous disease problems will be diagnosed in domestic chimpanzee colonies that are closely monitored. Miscellaneous disease problems that have been reported include oral and dental diseases [35, 36], alkaptonuria [37], diabetes [38], air sac infections [39], myocardial infarcts and cerebral aneurysms [40–42] and appendicitis [43].

Some of the more significant miscellaneous disease problems encountered in the Yerkes chimpanzee colony over the past 25 years include 6 animals with myocardial fibrosis, 2 animals with brain hemorrhage or infarcts associated with severe atherosclerosis, 2 cases of pulmonary thromboembolism, 1 case of diabetes mellitus, 1 cytogenetic abnormality comparable to Down's syndrome of humans [44], 1 hiatal hernia with obstruction of the esophagus and kidney lesions in 1 animal that were comparable to those seen with malignant hypertension in humans.

Summary and Conclusions

In this article, we have briefly reviewed the diseases that have been reported in chimpanzees and have reviewed the clinicopathologic data derived from the Yerkes chimpanzee colony over a 25-year period. During this period, necropsy examinations were done on 186 chimpanzees. Abortions, stillbirths and neonatal deaths accounted for almost 50% of the necropsies. Over half of the neonatal deaths occurred on the day of birth, with 75% of the neonatal deaths occurring within the first week of life. Trauma inflicted by other chimpanzees in the group accounted for almost 50% of the neonatal deaths.

In general, the most common causes of clinical disease problems are bacterial infections, with less frequent causes being viral or parasitic infections. Mycotic infections and neoplastic diseases have been infrequently reported in chimpanzees and occurred very infrequently in chimpanzees of the Yerkes colony. Our observations with respect to the most commonly encountered diseases and lesions in laboratory maintained chimpanzees are comparable to those reported for other domestic chimpanzee colonies [31, 32].

It is clear that a wide variety of clinical diseases and postmortem lesions will be encountered in closely monitored chimpanzees in research colonies. Detailed clinicopathologic evaluation of spontaneous diseases in chimpanzees is an important adjunct to the maintenance and use of this very important and unique laboratory animal species. Appropriate management practices, preventive medicine programs and clinicopathologic diagnostic support will have a major impact on the control of spontaneous diseases in domestic chimpanzee colonies. Based on our observations and that reported from other chimpanzee colonies, special efforts are needed to decrease the incidence of abortions, stillbirths and neonatal deaths in chimpanzee breeding colonies.

References

1 McClure HM, Strozier LM, Keeling ME: Enteropathogenic *Escherichia coli* infection in anthropoid apes. J Am Vet Med Assoc 1972;161:687–689.

2 McClure HM, Brodie AR, Anderson DC, Swenson RB: Bacterial infections of nonhuman primates; in Benirschke K (ed): Primates: The Road to Self-Sustaining Populations. New York, Springer, 1986; pp 531–556.

3 McClure HM, King FA: Yersiniosis: A review and report of an epizootic in nonhuman primates; in Ryder OR, Byrd ML (eds): One Medicine. Berlin, Springer, 1984, pp 217–241.

4 Blumberg HM, Kiehlbauch JA, Wachsmuth IK: Molecular epidemiology of *Yersinia enterocolitica* 0:3 infections: Use of chromosomal DNA restriction fragment length polymorphisms of rRNA genes. J Clin Microbiol 1991;29:2368–2374.

5 Donham KJ, Leininger JR: Spontaneous leprosy-like disease in a chimpanzee. J Infect Dis 1977; 136:132–136.

6 Gormus BJ, Xu K, Alford PL, Lee DR, Hubbard GB, Eichberg JW, Meyers WM: A serologic study of naturally acquired leprosy in chimpanzees. Int J Leprosy 1991;59:450–457.

7 Hubbard GB, Lee DR, Eichberg JW, Gormus BJ, Xu K, Meyers WM: Spontaneous leprosy in a chimpanzee *(Pan troglodytes)*. Vet Pathol 1991;28:546–548.

8 McClure HM, Guilloud NB: Comparative pathology of the chimpanzee; in Bourne GH (ed): The Chimpanzee. Basel, Karger, 1971, vol 4, pp 103–272.

9 McClure HM, Keeling ME: Viral diseases noted in the Yerkes Primate Center colony. Lab Anim Sci 1971;21:1001–1010.

10 McClure HM, Swenson RB, Kalter SS, Lester TL: Natural genital *Herpesvirus hominis* infection in chimpanzees *(Pan troglodytes* and *Pan paniscus)*. Lab Anim Sci 1980;30:895–901.

11 Van Ranst M, Fuse A, Sobis H, de Meurichy W, Syrjanen SM, Billiau A, Opdenakker G: A papillomavirus related to HPV type 13 in oral focal epithelial hyperplasia in the pygmy chimpanzee. J Oral Pathol Med 1991;20:325–331.

12 Blakeslee JR Jr, McClure HM, Anderson DC, Bauer RM, Huff LY, Olsen RG: Chronic fatal disease in gorillas seropositive for simian T-lymphotropic virus I antibodies. Cancer Lett 1987; 37:1–6.
13 Lanford RE, Notvall L, Barbosa LH, Eichberg JW: Evaluation of a chimpanzee colony for antibodies to hepatitis C virus. J Med Virol 1991;34:148–153.
14 Abe K, Inchauspe G. Shikata T, Prince AM: Three different patterns of hepatitis C virus infection in chimpanzees. Hepatology 1992;15:690–695.
15 Schlauder GG, Leverenz GJ, Amann CW, Lesniewski RR, Peterson DA: Detection of the hepatitis C virus genome in acute and chronic experimental infection in chimpanzees. J Clin Microbiol 1991;29:2175–2179.
16 Renquist DM, Johnson AJ, Lewis JC, Johnson DJ: A natural case of *Schistosoma mansoni* in the chimpanzee *(Pan troglodytes verus)*. Lab Anim Sci 1975;25:763–768.
17 Fremming BD, Vogel FS, Benson RE, Young RJ: A fatal case of amebiasis with liver abscesses and ulcerative colitis in a chimpanzee. J Am Vet Med Assoc 1955;126:406–407.
18 Miller MJ, Bray RS: *Entamoeba histolytica* infections in the chimpanzee *(Pan satyrus)*. J Parasitol 1966;52:386–388.
19 Holmes DD, Kosanke SD, White GL, Lemmon WB: Fatal enterobiasis in a chimpanzee. Am J Vet Res 1980;177:911–913.
20 Keeling, ME, McClure HM: Pneumococcal meningitis and fatal enterobiasis in a chimpanzee. Lab Anim Sci 1974;24:92–95.
21 McClure HM, Strozier LM, Keeling ME, Healy GR: Strongyloidosis in two infant orangutans. J Am Vet Med Assoc 1973;163:629–632.
22 McClure HM, Keeling ME, Custer RP, Marshak RR, Abt DA, Ferrer JF: Erythroleukemia in two infant chimpanzees fed milk from cows naturally infected with bovine C-type virus. Cancer Res 1974;34:2745–2757.
23 Chandler FW, McClure HM, Campbell WG Jr, Watts JC: Pulmonary pneumocystosis in nonhuman pimates. Arch Pathol Lab Med 1976;100:163–167.
24 Klokke AH, de Vries GA: Tinea capitis in chimpanzees caused by *Microsporum canis* Bodin 1902 resembling *M. obesum* Conant 1937. Sabouraudia 1963;2:268–270.
25 Otcenasek M, Dvorak J, Ladzianska K: *Trichophyton rubrum*-like dermatophyte as a causative agent of dermatophytosis in chimpanzees. Mycopathol Mycol Appl 1967;31:33–37.
26 Takashio M, de Vroey C: Piedra noire chez des chimpanzés du Zaïre. Sabouraudia 1975;13:58–62.
27 Roy AD: Rhinophycomycosis entomophthorae occurring in a chimpanzee in the wild in East Africa: Further report. Am J Trop Med Hyg 1974;23:935.
28 Roy AD, Cameron HM: Rhinophycomycosis entomophthorae occurring in a chimpanzee in the wild in East Africa. Am J Trop Med Hyg 1972;21:234–237.
29 Strutton WH: A case report of *Coccidioides immitis* in an immature chimpanzee. Aeromedical Research Laboratory Technical Report No 64–16. Holloman Air Force Base, 1964.
30 Schmidt RE, Butler TM: Esophageal candidiasis in a chimpanzee. J Am Vet Med Assoc 1970; 157:722–723.
31 Schmidt RE: Systemic pathology of chimpanzees. J Med Primatol 1978;7:274–318.
32 Hubbard GB, Lee DR, Eichberg JW: Diseases and pathology of chimpanzees at the Southwest Foundation for Biomedical Research. Am J Primatol 1991;24:273–282.
33 Muchmore E, Popper H, Peterson DA, Miller MF, Lieberman HM: Non-A, non-B hepatitis-related hepatocellular carcinoma in a chimpanzee. J Med Primatol 1988;17:235–246.
34 McClure HM: Neoplastic diseases in nonhuman primates: Literature review and observations in an autopsy series of 2,176 animals; in Montali RJ, Migaki G (eds): The Comparative Pathology of Zoo Animals. Washington, Smithsonian Institution Press, 1980, pp 549–565.
35 Schuman EL, Sognnaes RF: Developmental microscopic defects in the teeth of subhuman primates. Am J Phys Anthropol 1956;14:193–214.
36 Van Riper DC, Fineg J, Day PW, Douglas JD, Derwelis SK: Vincent's disease in the chimpanzee. J Am Vet Med Assoc 1967;151:905–906.
37 Watkins SP Jr, Binley H, Shulman NR: Alkaptonuria in a chimpanzee. 2nd Conf Exp Med Surg Primates, New York, 1969.
38 Rosenblum IY, Barbolt TA, Howard CF Jr: Diabetes mellitus in the chimpanzee *(Pan troglodytes)*. J Med Primatol 1981;10:93–101.
39 Strobert EA, Swenson RB: Treatment regimen for air sacculitis in the chimpanzee *(Pan troglodytes)*. Lab Anim Sci 1979;29:387–388.

40 Andrus SB, Portman OW, Riopelle AJ: Comparative studies of spontaneous and experimental atherosclerosis in primates. II. Lesions in chimpanzees including myocardial infarction and cerebral aneurysms. Progr Biochem Pharmacol 1968;4:393–419.
41 Manning GW: Coronary disease in the ape. Am Heart J 1942;23:719–724.
42 Stehbens WE: Cerebral aneurysms of animals other than man. J Pathol Bacteriol 1963;86:161–168.
43 Weinberg M: Un cas d'appendicite chez le chimpanzé. Ann Inst Pasteur 1904;18:323–331.
44 McClure HM, Belden KH, Pieper WA, Jacobson CB: Autosomal trisomy in a chimpanzee: Resemblance to Down's syndrome. Science 1969;165:1010–1012.

Harold M. McClure, DVM, Yerkes Regional Primate Research Center, Emory University, Atlanta, GA 30322 (USA)

Eder G, Kaiser E, King FA (eds): The Role of the Chimpanzee in Research.
Symp, Vienna 1992. Basel, Karger, 1994, pp 134–142

Comparison of Lipoproteins between Chimpanzees and Humans

Biochemical Aspects

Gerald Eder

Hans Popper Primate Center, Immuno AG, Clinical Research-Gastroenterology, Orth/Donau, Austria

Introduction

The composition of the individual lipoproteins differs in accordance with their location of synthesis or their function. A descriptive classification of hyperlipoproteinemia was introduced by Frederickson and Lees [6] on the basis of the electrophoretic mobility of lipoproteins. From that time onwards efforts aimed not only at improving Liebermann's description of the cholesterol reaction but mainly at quantitating the lipoproteins (VLDL, LDL, HDL) and the cholesterol distribution in these fractions as well as performing long-term studies for the evaluation of risk factors [7]. There is now an enormous amount of published papers on the relationship between lipids and coronary artery disease (CAD) in men. From the Framingham study and other studies carried out at the end of the 1960s it became clear that in large populations the concentration of the total cholesterol has a predictive value for the development of CAD [8, 9]. There is also a certain correlation between plasma triglycerides and CAD. However, it is still a matter of debate if triglyceride is an independent risk factor of coronary heart disease [10]. Data from 30 years follow up of the Framingham study have shown that there is not only the positive relationship between cholesterol and in particular LDL cholesterol but also a strong inverse relationship between HDL and coronary heart disease [11]. In the follow up of preventive studies it became now important to determine not only cholesterol, triglyceride and lipoproteins but in particular the apoprotein (Apo) content of the lipoproteins. In 1982 we started a study on the lipoproteins, cholesterol and triglycerides of chimpanzees of the Hans Popper Primate Center.

In the past 2 decades, chimpanzees have frequently been used in biomedical units for safety tests of blood derivatives and hepatitis vaccines as well as in AIDS research [1–3]. Yet, there are hardly any publications on comparative investigations regarding biochemical parameters and in particular the comparison with long-term studies in man. A typical example is the lipid metabolism. One might expect that lipoproteins in chimpanzees as the animal closest to humans would resemble the situation in humans. The influence of dietary fat and cholesterol on the lipoprotein profile of chimpanzees has been reported in 2 studies [4, 5]. The effects of a specific diet or elevated cholesterol values on the formation of atherosclerotic plaques have not been studied yet in chimpanzees.

Material and Methods

Chimpanzees

In 1982 24 juvenile chimpanzees were housed at the Hans Popper Primate Center. The animals were caught in the wild and imported either from the USA or from Africa. The chimpanzees were housed in individual cages and in animal rooms corresponding to biosafety level 3 [12]. The animals were fed mainly with fresh fruit (oranges, bananas, apples), vegetables as tomatoes, cabbage, carrots and tea, water ad libitum, full-cream milk and various milk-cereal formulas. Creative feeding was provided meaning that the food offered to each animal differed in the type of fruits and vegetables by week and by season. This scheme was not changed within the next 10 years. The group was investigated on their lipoprotein profile in 1982 and 1991/92. In addition, 27 other chimpanzees acquired within this decade were also tested for the same parameters in 1991/1992.

Blood Sampling

All blood samples were obtained in ketamine® anesthesia after a fasting period of 12 h. Blood was drawn under sterile conditions from a superficial cubital vein after putting an arm cuff to obtain compression between systolic and diastolic pressure. Serum was separated by centrifugation, and biochemical, immunological tests and lipoprotein electrophoresis were performed on the same day. In the series in 1982 on an average 15 samples of each of the 24 chimpanzees over a period of 6 months were investigated. In the series in 1991 and 1992, 7 samples of the same 24 animals, which were examined in 1982 and 15 samples of the 27 additionally acquired animals were analyzed. All blood samples were taken either on the occasion of safety testings of vaccines or on the occasion of a routine health check. It is our experience that safety tests of vaccines never influenced any of the parameters of this study.

Determination of Total Cholesterol

Total cholesterol was determined using commercially available reagent kits (Monotest® -cholesterol; Boehringer Mannheim, CHOD-PAP method) [13]. For internal quality control 3 commercially available control sera were used (Precinorm U®, Precinorm L® and Qualitrol®). For humans cholesterol values <200 mg/dl are regarded as normal.

Determination of Triglyceride

Triglycerides were hydrolyzed with subsequent determination of the liberated glycerol by colorimetry [14]. Commercially available reagents (triglyceride GPO-PAP, Boehringer Mannheim) were used. For internal quality control 3 different control sera were used (see above). Humans with triglyceride values of >200 mg/dl have an increased risk of CAD.

Determination of HDL-LDL-VLDL Cholesterol

In a first step lipoproteins were separated by electrophoresis in an agar medium containing human albumin. The lipoprotein fractions were visualized by chemical precipitation with polyanions [15, 16]. These lipoprotein bands were evaluated by a densitometer, and the HDL cholesterol, LDL cholesterol and VLDL cholesterol was calculated according to Wieland and Seidel [17]. Lipidophor All IN® (Immuno, Austria) was used for separation of the lipoprotein fractions by electrophoresis. In 1982, a specially adapted densitometer (Hirschmann Elscript®) and in 1992 a complete computerized system (Liposcript AT® Immuno) were used for evaluation and calculation of HDL cholesterol, LDL cholesterol and VLDL cholesterol. According to the European Atherosclerosis Society, a normal risk exists if LDL cholesterol is <130 mg/dl and HDL cholesterol is >55 mg/dl for males and >65 mg/dl for females. A study performed by Riesen et al. [18] demonstrated a range of 53 ± 12 mg/dl for HDL cholesterol, 144 ± 28 mg/dl for LDL cholesterol and 17 ± 4 mg/dl for VLDL cholesterol in humans [18].

Determination of Lp(a)

Lp(a) is a lipoprotein which consists of 80% LDL and has 1 (or perhaps 2) copies of an antigen [apo(a)] which is bound to Apo B-100 by disulfide bridges [19]. The concentration in humans may vary from 100 mg/dl to not detectable [20]. Several clinical trials demonstrated the significant correlation between elevated plasma levels (>30 mg/dl) and the risk of CAD [21, 22]. The antigen is separated by electrophoresis in agarose gel containing a monospecific antibody. The rocket-like immune precipitates are visualized by staining with a protein dye (amidoblack 0.4 mmol and Coomassie blue 0.12 mmol). The length of the precipitate is proportional to the concentration of Lp(a) [23]. As internal standard a human control serum with about 80 mg/dl Lp(a) was used in 3 different dilutions. All used reagents were produced by Immuno.

Determination of Apo AI, AII and B

Apo AI, AII and B were determined by radial immunodiffusion. This method is based on the homogeneous diffusion of an antigen in a gel layer containing a specific antibody [24]. The reaction between antigen and antibody results in a precipitation ring. The diameter of this ring is proportional to the antigen concentration. For all 3 Apo, commercially available reagent kits including quality control sera (Immuno) were used. A summary of normal values of Apo using different methods was published by Riesen et al. [18]. Cordova et al. [25] reported in healthy controls (n=20) the following normal ranges: Apo AI 90 ± 18, Apo AII 35 ± 6 and Apo B 75 ± 21 mg/dl using radial immunodiffusion.

Results

In the group of 24 chimpanzees a total cholesterol mean concentration of 199 mg/dl and a standard deviation of 33 mg/dl were calculated in 1982 (n = 543). 10 years later the same group showed a total cholesterol of 192 mg/dl (mean value) with a standard deviation of 32 mg/dl (n = 188 determinations; table 1). Testing the additionally acquired animals (n = 27) in 1992 did not

Table 1. Mean values and standard deviation of the lipoprotein profile (mg/dl) 1982/1992

	1982		1992			
			A		B	
	x̄	SD	x̄	SD	x̄	SD
Cholesterol	199 ±	33	192 ± 32		198 ± 36	
Triglycerides	157 ±	59	178 ± 78		169 ± 66	
HDL cholesterol	70 ±	18	65 ± 17		67 ± 20	
LDL cholesterol	104 ±	22	118 ± 23		120 ± 28	
VLDL cholesterol	24 ±	14	9 ± 6		11 ± 8	
Lp(a)	205 ±	126	91 ± 55		116 ± 69	
Apo AI	157 ±	40	178 ± 34		173 ± 35	
Apo AII	49 ±	14	47 ± 8		45 ± 8	
Apo B	96 ±	21	75 ± 18		73 ± 20	

The 1982 group consisted of 24 animals; A = same group as in 1982 (n = 24); B = larger group including the 24 animals tested already in 1982 (n = 51).

dramatically alter the mean value. For the whole group a total cholesterol concentration of 198 mg/dl as mean value and a standard deviation of 36 mg/dl (n = 403 determinations) was calculated. There is no correlation between the cholesterol value in 1982, 1992 and the gained body weight within this period of time (R^2 = 0.3824). In 15 animals a slight decrease in the total cholesterol concentration not exceeding more than 15 mg/dl was detected. In the remaining 9 animals there was an increase in the cholesterol concentration of the same magnitude. In general there was no correlation between ageing, higher cholesterol values or gaining body weight. However, in 1 animal cholesterol values were repeatedly higher than 300 mg/dl (320–360 mg/dl); in another animal with values ranging always at 250 mg/dl, once a total cholesterol of 328 mg/dl was found. 39 animals had cholesterol values between 200 and 300 mg/dl, the majority between 200 and 230 mg/dl.

The obtained triglyceride results were similar to the total cholesterol values. In 1982 a mean value of 157 mg/dl and a standard deviation of 59 mg/dl (n = 531 determinations) were calculated. In 1992 a slight increase up to 178 mg/dl of the mean value and a greater standard deviation (78 mg/dl) were found. Testing the additional 27 animals and calculation of a new mean value resulted in 169 ± 66 mg/dl (n=404 determinations; see table 1). In 8 chimpanzees triglyceride values were elevated above 300 mg/dl (up to 391 mg/dl). Again there is no correlation of these high triglyceride values to body weight or age. Similar results were obtained for HDL, LDL and VLDL cholesterol. There is a slight increase in LDL cholesterol within the 10-year observation period and a

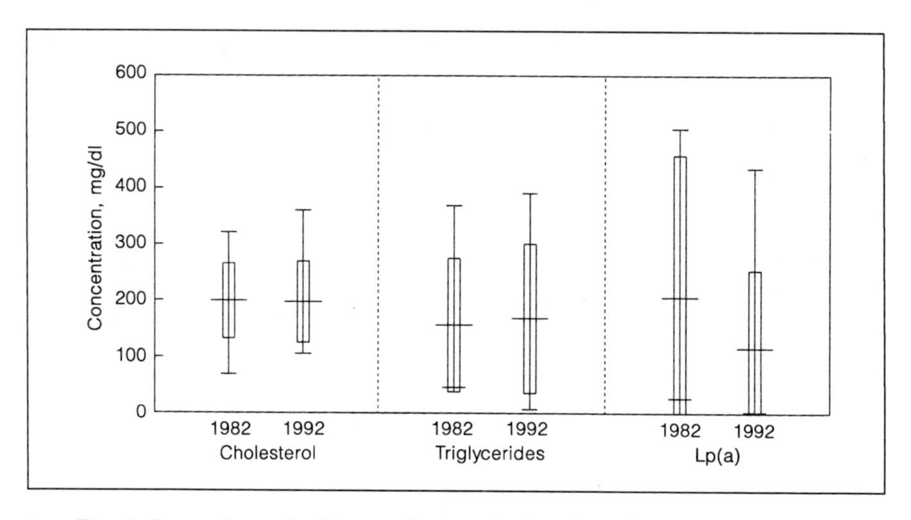

Fig. 1. Comparison of minimum (bottom line) and maximum values (top line) of cholesterol, triglycerides and Lp(a) 1982/1992; the horizontal middle line indicates the mean, the bar 2 standard deviations.

decrease in HDL cholesterol and a more pronounced decrease in VLDL cholesterol. All HDL cholesterol values were above 30 mg/dl. 18 of 51 animals (42 of 367 determinations) had LDL cholesterol values above 150 mg/dl, and 8 of 51 chimpanzees (14 of 367 determinations) showed VLDL cholesterol results above 30 mg/dl. The mean values of Apo AI, AII and B corresponded to these results (table 1). Apo AI was surprisingly high (mean 157 ± 40 mg/dl) with an increase in the mean value up to 178 mg/dl in 1992. Apo AII was as described for humans above 33 mg/dl with a mean value of 49 + 14 mg/dl. For Apo B a mean value of 96 mg/dl in 1982 and of 78 mg/dl and an SD of 18 mg/dl in 1992 were calculated. Surprising were the high Lp(a) concentrations in all of our chimpanzee. Table 1 shows a decrease in the mean Lp(a) concentration between 1982 and 1992. However, this phenomenon is very likely caused by the different techniques used in the 2 series. In 1982 we used a radial immunodiffusion technique with less sensitivity than the Laurell technique applied in 1992, but it remains a fact that all chimpanzees showed Lp(a) values above the upper limit for humans (>30 mg/dl) [21].

Discussion

In 2 series of testing, a normal range of 199 ± 33 mg/dl (mean ± SD) and of 198 ± 36 mg/dl total cholesterol could be established for healthy chimpanzees (fig. 1). This mean value is close to the risk limit for humans. In the USA it

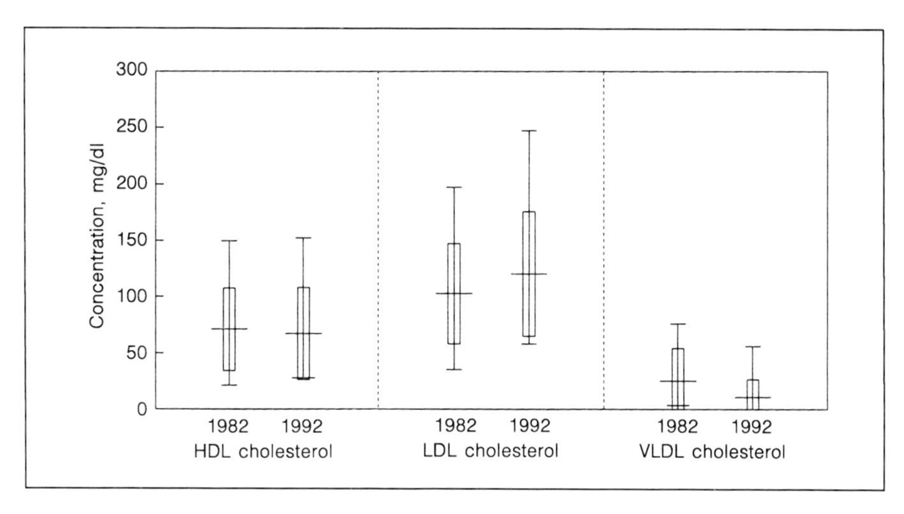

Fig. 2. Comparison of minimum and maximum values of HDL cholesterol, LDL cholesterol and VLDL cholesterol; for explanation, see figure 1.

is strongly suggested that patients with such a high cholesterol level should be given general education about coronary risk factors [26]. If those persons belong to the age group of 20–30 years such a cholesterol value is considered as moderate risk. Values of and above 220 mg/dl in this age group are indicative of high risk. It has to be pointed out that these risk factors apply only for the USA [27]. A person with the same cholesterol concentration (200–250 mg/dl) is classified as 'A' according to the management recommendations of the European Atherosclerosis Society [28]. It is recommended to assess the overall CAD risk, in those patients, to restrict food intake if overweight, to give nutritional advice and to correct other risk factors (smoking, high blood pressure). The blood pressure was monitored in our chimpanzee group over many years, and there was no correlation between high cholesterol or LDL cholesterol and high blood pressure. It is important to notice that in 8 chimpanzees the triglyceride values were above 300 mg/dl. Such values are considered in humans as indicative of a higher risk for CAD. In general there is a tendency of an increase in the triglyceride values in the tested chimpanzees. This phenomenon has to be carefully monitored to see if there is an influence from the food offered to the chimpanzees.

In 1982 and 1992, HDL cholesterol is on an average slightly higher in chimpanzees (fig. 2) than in humans, probably due to the difference in the HDL_2/HDL_3 ratio [4]. The Apo concentrations are in the range of their human counterparts [18]. Already in 1973, Scanu et al. [29] had demonstrated

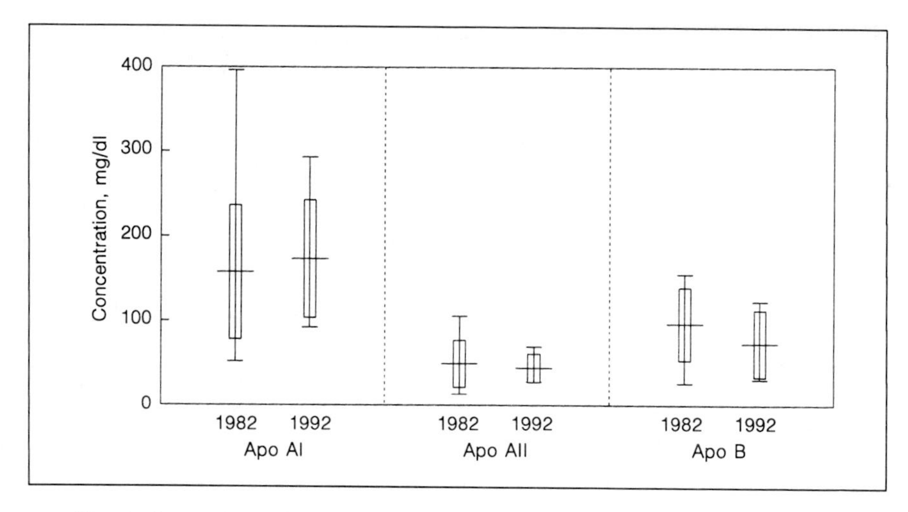

Fig. 3. Comparison of minimum and maximum values of Apo AI, AII and B; for explanation see figure 1.

in a single case study in a chimpanzee that the 2 major polypeptides Apo AI and Apo AII were different to those of rhesus macaques but similar to that of man in molecular weight as well as NH_2– and COOH– terminals. This was confirmed by our results, obtained by radial immunodiffusion using anti-human antisera (fig. 3). Concerning lipoprotein concentrations, mean values found in our study are also similar to those of human beings, which is in agreement with a study performed by Blaton and Peters [30] demonstrating that circulating plasma lipoproteins of chimpanzees are identical to man in composition as well as function. Using the human risk limits, we found 18 chimpanzees with LDL cholesterol values above 150 mg/dl [31]. The greatest difference was found in the concentration of Lp(a), which by far exceeded the risk limit for man in most chimpanzees. In humans a range between 0 and 150 mg/dl was established [32]. The majority of the tested patients without CAD risk was well below 30 mg/dl. High amounts of Lp(a) correlate with an increase in the risk of CAD in humans, assessed by angiography or flow measurements and complications such as myocardial infarction [33–35]. The role of this lipoprotein fraction in atherogenesis remains still unclear. At least 26 isoforms were described in connection with high and low plasma levels [36]. Cholesterol-rich nutrition seems not to influence the Lp(a) concentration [37]. The individual concentration is genetically determined and remains constant in humans for the whole life. Only in women a postmenopausal rise could be observed [20].

There is no doubt that the endothelium of blood vessels in the chimpanzee may show atherosclerotic alterations similar to man [30]. It remains to be established whether a chimpanzee with high Lp(a) is, like man, prone to the same risks of such alterations with all consequences or whether the increased concentration of Lp(a) is accompanied by or due to a different coagulation system.

References

1 Eder G, Bianchi L, Gudat F: Transmission of non-A, non-B hepatitis to chimpanzees: A second and third episode caused by the same inoculum; in Zuckerman AJ (ed): Viral Hepatitis and Liver Disease. Proceedings of the International Symposium on Viral Hepatitis and Liver Disease. New York, Liss, 1988, pp 550–552.
2 Tabor E, Purcell RH, Gerety RJ: Primate animal models and titered inocula for the study of human hepatitis A, hepatitis B, and hepatitis non-A, non-B. J Med Primatol 1983;12:305–318.
3 Barret N, Eder G, Dorner F: Characterisation of a vaccinia-derived recombinant HIV-1 gp160 candidate vaccine and its immunogenicity in chimpanzees. Biotechnol Ther 1991;2:91–106.
4 Blaton V, Vercaemst R, Vandecasteele H, Caster H, Peeters H: Isolation and partial characterization of chimpanzee plasma high density lipoproteins and their apoproteins. Biochemistry 1974;13:1127–1135.
5 Rosseneu M, Declercq B, Vandamme D, Vercaenst R, Soeteway F, Peeters H, Blaton V: Influence of oral polyunsaturated and saturated phospholipid treatment on the lipid composition and fatty acid profile of chimpanzee lipoproteins. Atherosclerosis 1979;32:141–153.
6 Fredrickson DS, Lees RS: System for phenotyping hyperlipoproteinemia. Circulation 1978; 31:321–327.
7 Wieland H, Seidl D: Fortschritte in der Analytik des Lipoproteinmusters. Inn Med 1978;5:290–300.
8 Kannel WB, Castelli WP, Gordon T, McNamara PM: Serum cholesterol, lipoproteins and the risk of coronary heart disease. Ann Intern Med 1971;74:1–12.
9 Epstein FH, Krueger DE: The changing incidence of coronary heart disease; in Jones M (ed): Modern Trends in Cardiology. London, Butterworths, 1969, pp 17–35.
10 Consensus conference: Treatment of hypertriglyceridemia. JAMA 1984;251:1196–2000.
11 Anderson KM, Castelli WP, Levy D: Cholesterol and mortality: 30 years of follow-up from the Framingham study. JAMA 1987;257:2176–2180.
12 Richardson JH, Barkley WE (eds): Biosafety in Microbiological and Biomedical Laboratories, ed 2. HHS Publ No (CDC) 84–8395. US Department of Health and Human Services, Public Health Service, Center for Disease Control and National Institutes of Health, 1988.
13 Siedel J, Haegele EO, Ziegenhorn J, Wahlefeld AW: Reagent for the enzymatic determination of serum total cholesterol with improved lipolytic efficiency. Clin Chem 1983;29:1075.
14 Bergmeyer HU: Methoden der enzymatischen Analyse, ed 3. Weinheim, Verlag Chemie, 1974, vol II, p 1878.
15 Seidel D, Wieland H, Ruppert C: Improved techniques for assessment of plasma lipoprotein patterns. I. Precipitation in gels after electrophoresis with polyanionic compounds. Clin Chem 1973;19:737–739.
16 Wieland H, Seidel D: Improved techniques for assessment of serum lipoprotein patterns. II. Rapid method for diagnosis of type III hyperlipoproteinemia without ultracentrifugation. Clin Chem 1973;19:1139–1141.
17 Wieland H, Seidel D: Fortschritte in der Analytik des Lipoproteinmusters. Inn Med 1978;5:290–300.
18 Riesen W, Mordasini R, Oster P: Ergebnisse der chemischen und immunologischen Bestimmung der Serumlipoproteine bei Hyperlipoproteinaemien. Lab Med 1981;5:103–111.
19 Houlston R, Friedl W: Biochemistry and clinical significance of lipoprotein (a.) Ann Clin Biochem 1988;25:499–503.
20 Sandkamp M, Assmann G: Lipoprotein (a) in PROCAM participants and young myocardial infarction survivers; in Scanu AM (ed): Lipoprotein(a): 25 Years of Progress. New York, Academic Press, 1990, pp 205–209.

21 Kostner GM, Avogaro P, Cazzolato G, Marth E, Bittolo-Bon G, Quinci GB: Lipoprotein Lp(a) and the risk for myocardial infarction. Atherosclerosis 1981;38:51–61.
22 Rhoads GG, Dahlen G, Berg K, Mortyon NE, Dannenberg AL: Lp(a) lipoprotein as a risk factor for myocardial infarction. JAMA 1986;256:2540–2544.
23 Kostner GM, Gries A, Pometta M, Molinari E, Pichler P, Aicher H: Immunochemical determination of lipoprotein Lp(a): Comparison of Laurell electrophoresis and enzyme-linked immunosorbent assay. Clin Chim Acta 1990;188:187–192.
24 Mancini G, Carbonara AO, Heremans JF: Immunochemical quantitation of antigens by single radial immunodiffusion. Immunochemistry 1965;2:235–254.
25 Cordova C, Musca A, Violi F, Alessandri C, Iuliano L: Apolipoproteins A-I, A-II and B in chronic active hepatitis and in liver cirrhotic patients. Clin Chim Acta 1984;137:61–66.
26 Consensus conference (USA): Lowering blood cholesterol to prevent heart disease. JAMA 1985; 253:2080–2086.
27 National cholesterol education programme (USA). Arch Int Med 1988;148:36–39.
28 European Atherosclerosis Society: Guidelines of study group. Eur Heart J 1987;8:77–88.
29 Scanu AM, Edelstein C, Wolf RH: Chimpanzee (*Pan Troglodytes*) serum high density lipoproteins: Isolation and properties of their 2 major apolipoproteins. Biochim Biophys Acta 1974;351:341–347.
30 Blaton V, Peeters H: The nonhuman primates as models for studying human atherosclerosis: Studies on the chimpanzee, the baboon and the rhesus macaque. Adv Exp Med Biol 1974;67:33–64.
31 Cremer P, Nagel D: Diagnostische Strategien zur Beurteilung von Fettstoffwechselstörungen und zur therapeutischen Zielsetzung. Internist 1992;33:32–37.
32 Dahlen GH, Guyton JR, Attar M, Farmer JA, Kautz JA, Gotto AM Jr: Association of levels of lipoprotein Lp(a), plasma lipids and other lipoproteins with coronary artery disease documented by angiography. Circulation 1986;74:758–765.
33 Armstrong VW, Cremer P, Eberle E, Manke A, Schulze F, Wieland H, Kreuzer H, Seidel D: The association between serum Lp(a) concentrations and angiographically assessed coronary atherosclerosis: Dependence on serum LDL levels. Atherosclerosis 1986;62:249–257.
34 Hearn JA, DeMaio SJ, Roubin GS, Hammarstrom M, Sgoutas D: Predictive value of lipoprotein(a) and other serum lipoproteins in the angiographic diagnosis of coronary artery disease. Am J Cardiol 1990;66:1176–1180.
35 Genest J, Jenner JL, McNamara JR, Ordovas JM, Silberman SR, Wilson PWF, Schaefer EJ: Prevalence of lipoprotein(a) [Lp(a)] excess in coronary artery disease. Am J Cardiol 1991;67:1039–1045.
36 Utermann G: The mysteries of lipoprotein(a). Science 1989;246:904–910.
37 Morrisett JD, Guyton JR, Gaubatz JW, Gotto AM Jr: Lipoprotein (a): Structure, Metabolism and epidemiology; in Gotto AM Jr (ed): Plasma Lipoproteins. Amsterdam, Elsevier, 1987, pp 129–152.

Gerald Eder, MD, Hans Popper Primate Center, Head, Clinical Research-Gastroenterology, Immuno AG, Uferstrasse 15, A–2304 Orth/Donau (Austria)

Eder G, Kaiser E, King FA (eds): The Role of the Chimpanzee in Research.
Symp, Vienna 1992. Basel, Karger, 1994, pp 143–153

The Chimpanzee in the Development of Vaccines for Parasitological Diseases

Special Reference to Malaria and Onchocerciasis

M.L. Eberhard[a], *D.A. Abraham*[b]

[a] Division of Parasitic Diseases, Centers for Disease Control and Prevention, Public Health Service, US Department of Health and Human Services, Atlanta, Ga., USA;
[b] Department of Microbiology and Immunology, Thomas Jefferson University, Philadelphia, Pa., USA

Historically, tropical areas of the world, with over one half of the population of the world, have suffered most from parasitic infections. Most tropical areas remain under the curse of heavy and intense parasitic infections, but we are also beginning to realize an ever growing increase in parasitic infections in temperate areas of the world. These have been most marked in day care and other child care facilities and in the HIV arena, where a number of the opportunistic infections responsible for many deaths in AIDS patients are parasitic in nature. These events, coupled with an ever growing movement of people to and from the tropics into temperate areas, have created a measurable increase in the number of laboratories actively studying parasitic infections. In the search for solutions to problems surrounding the control and eradication of parasitic diseases, laboratory animals, including and especially primates, continue to provide needed systems in which to study complex host-parasite systems.

Parasites, like other infectious organisms, can exhibit a wide range in host specificity, but in general, many tend to be fastidious in their choice of suitable natural or experimental hosts. Furthermore, because parasites are complex organisms, much more so than bacteria or viruses, many have not been adaptable to in vitro cultivation systems or other means of propagating and studying them outside the animal host. Therefore, the need for good animal models remains acute for the study of many parasitic infections. Chimpanzees have long held the interest of parasitologists because in nature they frequently

Table 1. Medically important human parasites reported as natural infections in chimpanzees

Protozoa	Helminths
Giardia	Hookworms
Entamoeba histolytica	*Ascaris*
Toxoplasma	*Trichuris*
Trypanosoma	*Strongyloides*
Plasmodium	*Schistosoma*
	Mansonella

Table 2. Plasmodium species known to infect man and chimpanzees, dates when each species was described and timing of erythrocytic cycle

	Man	Chimpanzee
Malignant tertian (36–48 h)	*P. falciparum* (1897)	*P. reichenowi* (1920)
Quartan (48 h)	*P. malariae* (1892)	*P. rodhaini* (1939)
Tertian (48–50 h)	*P. vivax* (1890) *P. ovale* (1922)	*P. schwetzi* (1920)

harbor species identical to, or indistinguishable from, those recognized in man (table 1). Other parasites may be almost exclusively found in man but can be experimentally established in chimpanzees.

In this overview, we will restrict our comments to two serious human pathogens universally recognized for the serious threat to life and community growth that they pose and to the role chimpanzees have played in our study and ultimate understanding and control of these diseases. The first parasitic infection to be discussed is malaria – probably the No. 1 killer of children in the tropics today. It is estimated that nearly 10% of the world's population is affected by malaria and between 1 and 2 million children die each year due to the infection. Malaria remains today as great or greater a problem than ever before due to factors such as increasing populations in endemic areas, shrinking health dollars and diverse problems including issues such as civil strife, famine and droughts. Drug resistance, on the part of both the parasite and the mosquito vector, has led to vast epidemics in areas once thought to be under control, including most areas of Southeast Asia and sub-Saharan Africa. Interestingly, malaria parasites indistinguishable from the four major human species were recognized years ago in chimpanzees (table 2) [1]. Questions concerning the exact taxonomic designation of these species are still problemat-

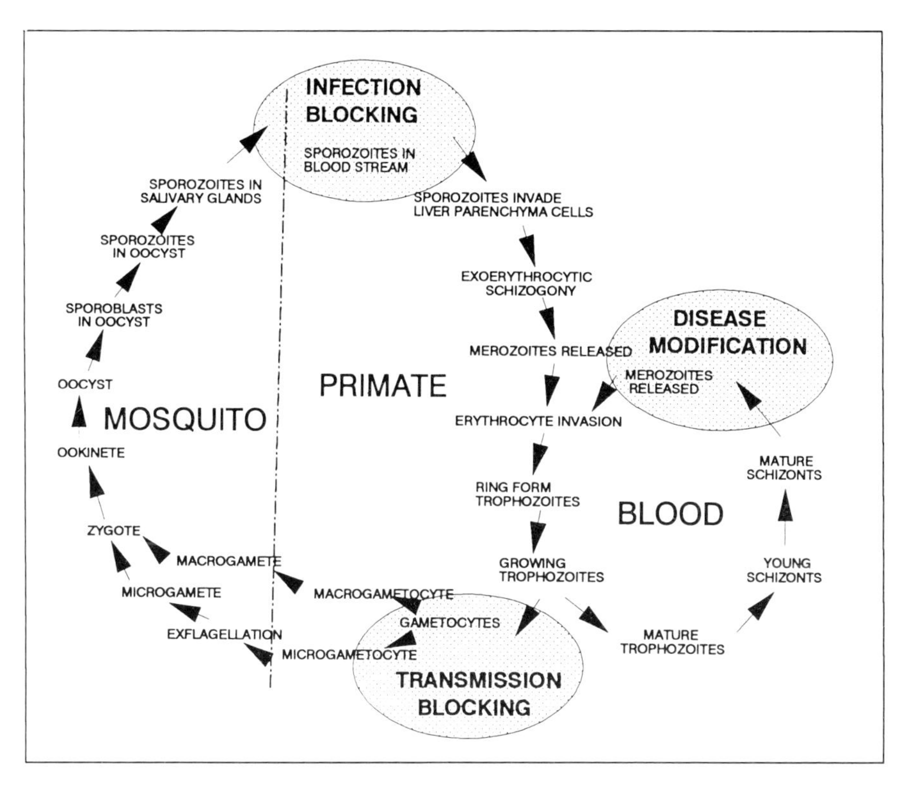

Fig. 1. Schematic representation of the life cycle in *Plasmodium* spp. depicting at least three places in the cycle where malaria vaccines may play a role in controlling this disease.

ic and the malaria parasites identified in the blood of chimpanzees retain their own separate names. They are, however, almost always referred to as the equivalent of one of the human species. These species may have shared common vectors and had a free exchange of parasites between man and chimpanzee hosts in the not too distant past. However, recent studies on cross-infection attempts and comparison of parasite antigens suggest that the primate malarias may indeed be distinct species [2, 3].

Chimpanzees have received considerable interest as experimental hosts of malaria and they have provided much needed answers about the biology and immunology of malaria in general. However, the role that chimpanzees have played in the development of malaria vaccines has been limited and for that reason comments will be restricted to a very brief overview of the status of malaria vaccines. To understand malaria vaccine research, one has to appreciate the life cycle (fig. 1). The malaria life cycle is complex, but because of its complexity, there are at least three different areas which can be attacked from the point of view of developing a vaccine. The first possible effect, directed at the sporozoite, would be prevention of, or reduction in, infection upon exposure. This would be an infection-blocking vaccine. A second potential vaccine effect would be directed at modifying the disease by either reducing the severity or shortening the duration of illness. This would be a disease-modifying vaccine. A third area of vaccine study encompasses blocking or reducing transmission of the parasite to the mosquito so that people would be less infectious or infectious for a shorter time. This would constitute a transmission-blocking vaccine. These are stage-specific vaccines and most vaccines of this type developed to date have been 'leaky', that is they have not induced complete protection. Whether candidate vaccines are immunogenic or actually confer protection has been the subject of some debate [4]. It is also unknown whether they may be sensitive to boosting by natural infection or exposure which could either prolong the immunity or intensify the immunity [5]. Despite widespread anticipation of a malaria vaccine in recent years [6], the actual development of a workable product has not occurred. As a result, more realistic goals have been set for malaria vaccines which include prevention of death, anemia and cerebral malaria rather than total prevention of infection. However, intense research activity continues, and as an aid to the development of rational approaches to vaccination, extensive mathematical modeling of the transmission dynamics of malaria is currently being refined [7]. On the more practical side, lower primates such as squirrel monkeys will continue to be the cornerstone for vaccine trials. Chimpanzees will probably not be used extensively because of cost, availability and housing requirements associated with keeping the large numbers of animals required for these studies.

The second disease for consideration is onchocerciasis, another devastating parasitic infection of man in tropical Africa and Central and South America, which affects some 20–30 million persons. River blindness derives its name from the fact that the insect vector, a small biting blackfly, breeds in fast-flowing rivers and streams and the most serious sequela of infection is blindness. Onchocerciasis is so serious in some areas that it has led to the movement of entire villages away from fertile, productive riverine land into areas less suitable for sustaining a living just to escape the skin and eye lesions associated with infection. As with malaria, the life cycle is complex and has several stages which may be appropriate to attack with vaccines (fig. 2). *Onchocerca volvulus* has both insect and a vertebrate stages in the life cycle. A vaccine directed at the microfilaria stage would be transmission blocking in nature, whereas a vaccine against infective larvae or very immature worms would be infection blocking. A vaccine against the microfilariae might also be disease modifying in action.

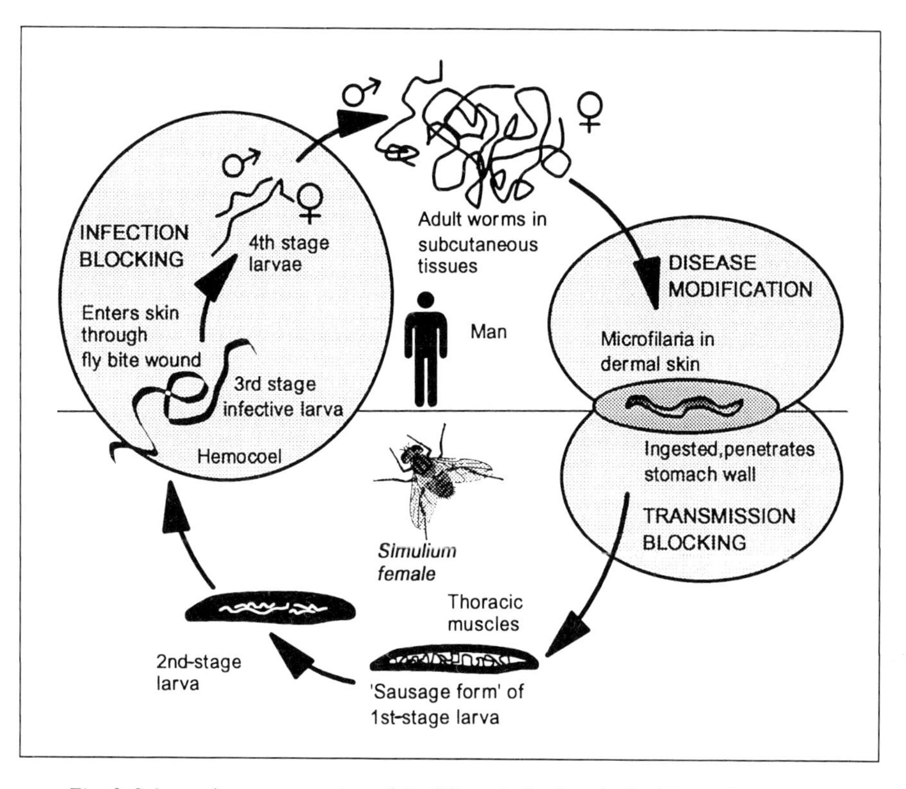

Fig. 2. Schematic representation of the life cycle in *O. volvulus* illustrating several areas where potential vaccines could be used to attack this disease.

In contrast to the situation in malaria, bringing potential vaccine candidates to trial has proceeded much slower in onchocerciasis. This may be due, in part, to the nature of the beast. *Onchocerca* cannot be grown in vitro, animals cannot be infected simply by transfusing blood from an infected animal into another like in malaria, chimpanzees are the only recognized laboratory host [8] and the prepatent period (time from inoculation till production of microfilariae) is 9–18 months [9]. Nevertheless, considerable research has focused on protective immunity, antigen selection and antigen production and development of animal screening systems. Also in contrast to malaria vaccine studies, chimpanzees have already played a crucial role in understanding the hosts' immunological response [10–12] and will undoubtedly continue to play a key role in vaccine development and testing [13]. In order to better conserve both the use of scarce parasite material and valuable chimpanzees, a screening system has been developed for testing antigens for potential as vaccines

Table 3. Proposed stages for screening *Onchocerca* antigens

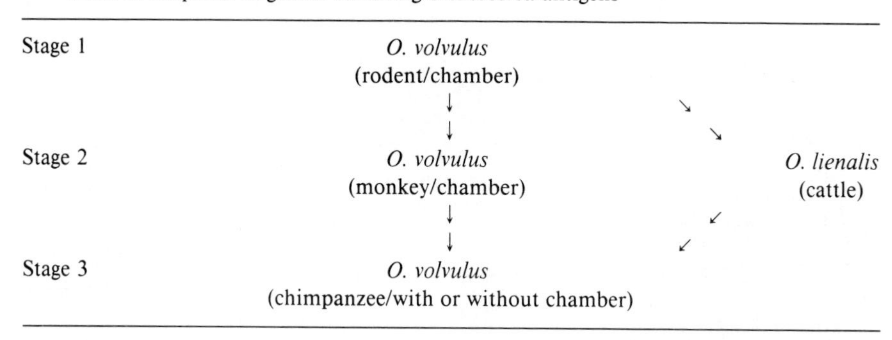

Stage		
Stage 1	*O. volvulus* (rodent/chamber) ↓ ↓	↘ ↘
Stage 2	*O. volvulus* (monkey/chamber) ↓ ↓	*O. lienalis* (cattle) ↙ ↙
Stage 3	*O. volvulus* (chimpanzee/with or without chamber)	

(table 3). Stage 1 screening would be accomplished in rodents with *O. volvulus* and/or related animal species of *Onchocerca*. If results are encouraging, then tests would progress to stage 2 with either the *O. volvulus*–monkey/chamber system or *Onchocerca* spp. in cattle. Stage 3 testing would be in the *O. volvulus*–chimpanzee model. Although chimpanzees are the only experimental host for *O. volvulus*, the use of implantable diffusion chambers has allowed us and others to use surrogate rodent and lower primate hosts to study the early growth, development and survival of *Onchocerca* larvae and conduct vaccine screening [14–17]. The use of rodents clearly expands the capability of screening many agents which would not be feasible if we were restricted to the use of chimpanzees. Chambers are constructed from Lucite rings and covered with filter membranes. These membranes can be either cell excluding if small pore size membranes are used (0.1–0.5 μm) or permit cell entry if a larger pore size is used (not so large as to permit escape of larvae). Third-stage larvae are injected into the chamber (fig. 3), sealed and then implanted subcutaneously into the animal (fig. 4, 5). The advantage of using rodents, as stated, is the low cost of doing large numbers of trials with compounds of unknown value. The disadvantage is that only a single chamber per animal is feasible. The distinct advantage, then, of working with chimpanzees, is the ability to implant 15–20 chambers at one time and be able to remove a few at predetermined times and have your whole experiment run in a single animal. The advantages of using chambers are obvious. The greatest problem which we faced was to convince not only ourselves but the scientific community that results obtained from chamber implants in any animal (rodent or primate) were valid tests and that results from rodents were consistent with what was happening in the chimpanzee system. We looked at three main criteria to evaluate the use of chambers, including survival of the larvae, growth of the larvae and, most crucially, whether the larvae molted from third- to fourth-stage larvae. To evaluate these

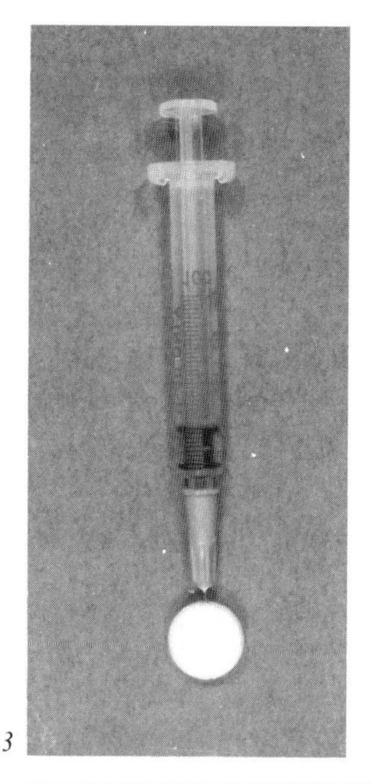

3

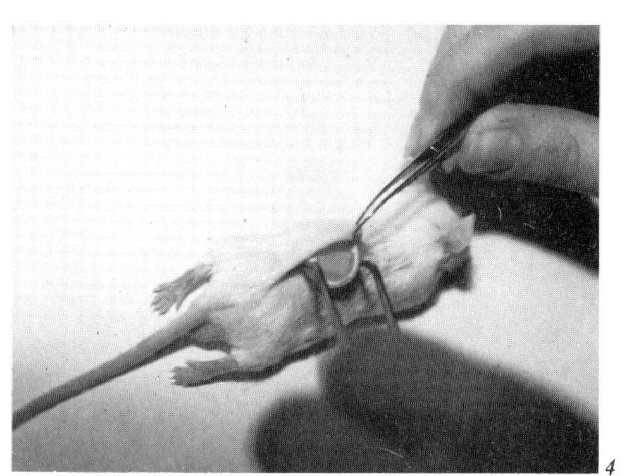

4

5

Fig. 3. Assembled chamber being loaded with *Onchocerca* larvae via needle and syringe.

Fig. 4. Charged and sealed chamber being inserted into a subcutaneous pocket on the back of a mouse.

Fig. 5. Mice, each harboring a single chamber which can be removed at predetermined times to monitor the survival and growth of larvae.

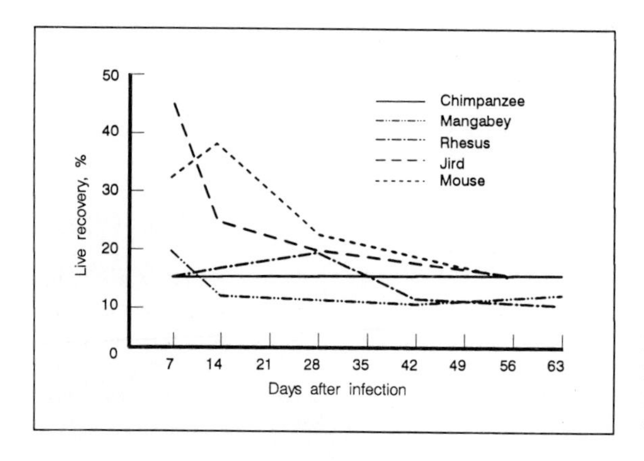

Fig. 6. Survival of *O. volvulus* larvae, depicted as mean percent recovery, after implantation into primates and rodents over 7- to 63-day intervals.

systems we set up a large series of experiments comparing the outcome of chambers implanted in a variety of rodents, lower primates and chimpanzees. Results obtained from chambers implanted into chimpanzees were always viewed as the 'gold standard' to which other systems had to measure up.

Very briefly, larval survival in chambers implanted into rodents was comparable to that seen in primates, including chimpanzees (fig. 6). Actually, for the first week, a greater proportion of larvae could be recovered from chambers implanted into rodents. However, by 14 days recovery from all hosts was about equal, and at the longest time intervals tested, survival was slightly better in chimpanzees. Larvae grew appreciably over the first 14–28 days in all hosts tested and larvae implanted into rodents achieved mean maximal lengths equivalent to that seen in larvae implanted into primate hosts (fig. 7). Molting of third-stage to fourth-stage larvae was observed to commence as early as day 3 after implantation and by day 7, 50–100% of live larvae were observed to have successfully completed the molt. By 14 days of implantation, molting rates were not significantly different for primate or rodent hosts (fig. 8). These studies have demonstrated that chambers provide a very practical method for the study of onchocerciasis under laboratory conditions. Further, we have been able to outline a logical succession of experimental testing from rodent to lower primate through chimpanzees.

It is important to remember several points about what we can and cannot do with the use of chambers. They have allowed us to study the early development of the parasite that had not been possible, they allow us to hold elusive worms 'captive' so that we can find and recover them at will and they allow us to directly measure the outcome of an experiment without sacrificing the animal. We must assume, however, that the behavior, survival and growth of the parasite in chambers may not duplicate exactly how the parasite would

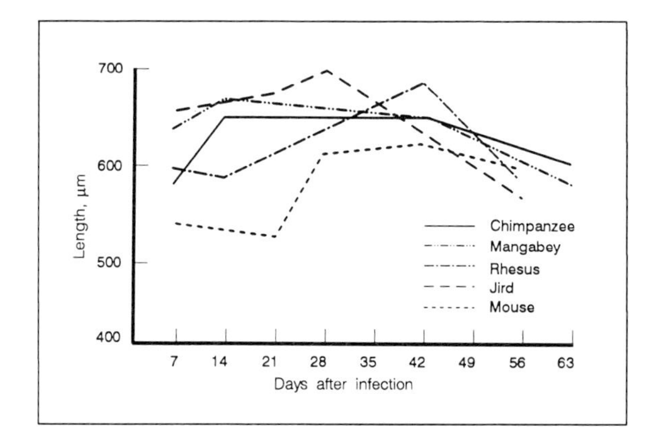

Fig. 7. Growth of *O. volvulus* larvae, depicted as mean lengths, after implantation into primates and rodents over 7- to 63-day intervals.

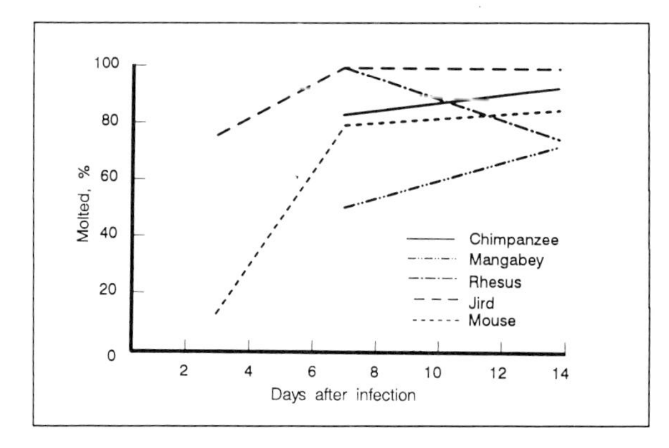

Fig. 8. Rate of molting in *O. volvulus* larvae, depicted as percent molted, after implantation into primates and rodents over 3- to 14-day intervals.

normally behave if it could freely wander in the hosts' tissues. Clearly, fluids, nutrients or cells may pass into and out of the chamber, but how important it may be for the larvae to be in direct contact with the host's tissues is unknown. Therefore we view the use of chambers as a valuable tool to evaluate drug or vaccine trials, but we recognize that there are limitations to the system and we are prepared to work within certain constraints.

It should be evident then that chimpanzees will continue to play a crucial role in the future development of vaccines against onchocerciasis. Should it ultimately prove possible to combat and reduce suffering from this devastating disease by the use of vaccines and if we are able to save the eyesight of future generations, mankind will again be indebted to primates, especially chimpanzees, for their contribution to scientific discovery and our well being.

Conclusion

The major health impact that malaria has in the tropics today has directed more effort toward developing a vaccine(s) for malaria than for any other parasitic infection. Potential malaria vaccines, targeting specific areas in the life cycle, could be infection blocking, disease modifying or transmission blocking in nature. Although considerable information about the biology of malaria was provided by the study of chimpanzees, owl monkeys (*Aotus* sp.) and squirrel monkeys (*Saimiri* sp.) have now supplanted chimpanzees as the major primate model for vaccine trials. In contrast, vaccine development for onchocerciasis has proceeded slowly, but because of their unique role as an experimental host for *Onchocerca*, chimpanzees have played a critical role in the laboratory research on onchocerciasis. Recently, a screening system using implantable diffusion chambers has been developed for testing antigens for potential vaccine trials in both surrogate hosts (such as rodents and lower primates) and chimpanzees. Results obtained from chambers implanted into chimpanzees provided the standard by which other systems were compared and evaluated and provided the conceptual framework for a three-tiered screening system. The use of chambers has provided an alternate means for evaluating potential vaccines in an in vivo system, and it appears that chimpanzees will continue to play a crucial role in the development of a vaccine against onchocerciasis.

References

1 Coatney GR, Collins WE, Warren M, Contacos PG: The Primate Malarias. Washington, US Printing Office, 1971.
2 Lal AA, Goldman IF, Campbell GH: Primary structure of the 25-kilodalton ookinete antigen from *Plasmodium reichenowi*. Mol Biochem Parasitol 1990;43:143–146.
3 Lal AA, Goldman IF: Circumsporozoite protein gene from *Plasmodium reichenowi*, a chimpanzee malaria parasite evolutionarily related to the human malaria parasite *Plasmodium falciparum*. J Biol Chem 1991;266:6686–6689.
4 Targett GAT: SPf66, a candidate synthetic malaria vaccine: Immunogenicity versus protection. Parasitol Today 1992;8:354–355.
5 Deloran P, Chougnet C. Is immunity to malaria really short-lived? Parasitol Today 1992;8:375–378.
6 McGregor I: Clinical trials of new malaria vaccines. Parasitol Today 1992;1:32–33.
7 Halloran ME, Struchiner CJ: Modeling transmission dynamics of stage-specific malaria vaccines. Parasitol Today 1992;8:77–85.
8 Duke BOL: Experimental transmission of *Onchocerca volvulus* from man to a chimpanzee. Trans R Soc Trop Med Hyg 1962;56:271.
9 Duke BOL: Observations on *Onchocerca volvulus* in experimentally infected chimpanzees. Tropenmed Parasitol 1980;31:41–54.
10 Weiss N, van den Ende MC, Albiez EJ, Barbiero VK, Forsyth K, Prince AM: Detection of serum antibodies and circulating antigens in a chimpanzee experimentally infected with *Onchocerca volvulus*. Trans R Soc Trop Med Hyg 1986;80:587–591.

11 Soboslay PT, Dreweck C, Taylor HR, Wenk P, Greene BM: Immune response in chimpanzees experimentally infected with *Onchocerca volvulus*. Tropenmed Parasitol 1987;38:342.
12 Eberhard ML, Dickerson JW, Boyer AE, Tsang VCW, Zea-Flores R, Walker EM, Richards FOR, Zea-Flores G, Strobert E: Experimental *Onchocerca volvulus* infections in mangabey monkeys *(Cercocebus atys)* compared to infections in human and chimpanzees *(Pan troglodytes)*. Am J Trop Med Hyg 1991;44:151–160.
13 Prince AM, Brotman B, Johnson EH Jr, Smith A, Pascual D, Lustigman S: *Onchocerca volvulus*: Immunization of chimpanzees with X-irradiated third-stage (L3) larvae. Exp Parasitol 1992;74: 239–250.
14 Strote G: Development of infective larvae of *Onchocerca volvulus* in diffusion chambers implanted into *Mastomys natalensis*. Tropenmed Parasitol 1985;36:120–122.
15 Bianco AE, Mustafa MB, Ham PJ: Fate of developing larvae of *Onchocerca lienalis* and *O. volvulus* in micropore chambers implanted into laboratory hosts. J Helminthol 1989;63:218–226.
16 Abraham D, Eberhard ML, Lange AM, Yutanawiboonchi W, Perler FB, Lok JB; Identification of surrogate rodent hosts for larval *Onchocerca volvulus* and induction of protective immunity in a rodent model. J Parasitol 1992;78:447–453.
17 Abraham D, Lange AM, Yutanawiboonchai, Trpis M, Dickerson JW, Swenson B, Eberhard ML: Survival and development of larval *Onchocerca volvulus* in diffusion chambers implanted in primate and rodent hosts. J Parasitol 1993;79:571–582.

Mark L. Eberhard, Division of Parasitic Diseases F13, National Center for Infectious Diseases, Centers for Disease Control, 1600 Clifton Road, Atlanta, GA 30333 (USA)

Eder G, Kaiser E, King FA (eds): The Role of the Chimpanzee in Research.
Symp, Vienna 1992. Basel, Karger, 1994, pp 154–155

Testing of Recombinant Vaccines in Chimpanzees

F. Dorner, G. Antoine, F.G. Falkner

Immuno AG, Research Center, Orth/Donau, Austria

Successful immunization against infectious disease often requires a multicomponent host immune response against a variety of antigenic determinants. These responses include non-specific mechanisms, such as activation of the reticulo-endothelial system or specific responses, such as induction of specific antibody and effector T cell populations. Although parenteral immunization with whole, inactivated organisms or purified subunit vaccines results primarily in the synthesis of serum antibody, the ability to induce cell-mediated effector responses to specific antigenic determinants has been limited to immunization with live micro-organisms. Two methods of vaccine delivery, providing alternatives to vaccination with inactivated organisms or purified subunits in order to stimulate a broader and perhaps more effective host response, are used in our laboratories for several fields of application. Immunization by ingestion of *Salmonella* via the oral route results in induction of both humoral and cell-mediated immune responses. We and several other investigators have utilized attenuated strains of *Salmonella* harbouring plasmids encoding foreign antigens to deliver these antigens to the immune systems, resulting in the induction of specific systemic and mucosal responses [1, 2]. Different recombinant *Salmonella* strains (e.g. expression of heterologous epitopes as recombinant flagella on the surface of attenuated *Salmonella*) will be presented and there will be a discussion about the importance of the use of chimpanzees versus the mouse model, which is not directly applicable to human immunization, because non-typhoidal *Salmonella* species do not usually translocate to the mesenteric lymphatics in humans. The live vector technology has also been applied in our laboratories to fowlpox virus. We are currently investigating whether a chimaeric fowlpox virus can induce protective immunity in a non-avian host. The *env* gene of HIV-1 has been introduced by direct molecular cloning [3] into a chimaeric fowlpox virus. The cloning strategy and the use of chimpanzees as a model for the evaluation of the protective efficacy of chimaeric poxviruses is of great importance.

References

1 Curtiss R III, Kelly SM: *Salmonella typhimurium* deletion mutants lacking adenylate cyclase and cyclic AMP receptor protein are avirulent and immunogenic. Infect Immun 1987;55:3035–3043.

2 Dougan G, Chatfield S, Pickard D, Bester J, Callaghan DO, Maskell D: Construction and characterization of vaccines strains of Salmonella harboring mutations in two different *aro* genes. J Infect Dis 1988;158:1329–1335.

3 Scheiflinger F, Dorner F, Falkner FG: Construction of chimeric vaccinia viruses by molecular cloning and packaging. Proc Natl Acad Sci USA 1992;89:9977–9981.

Friedrich Dorner, PhD, Immuno AG, Research Center,
Uferstrasse 15, A-2304 Orth/Donau (Austria)

Eder G, Kaiser E, King FA (eds): The Role of the Chimpanzee in Research.
Symp, Vienna 1992. Basel, Karger, 1994, pp 156–165

Safety Testing of Blood Products in Chimpanzees

Gerald Eder[a], *Ildiko Sarosi*[b], *Zsuzsa Schaff*[b]

[a] Hans Popper Primate Center, Immuno AG, Clinical Research – Gastroenterology, Orth/Donau, Austria;
[b] Institute of Pathology and Experimental Cancer Research, Semmelweis Medical University, Budapest, Hungary

Introduction

In 3 clinical studies cases of hepatitis B and hepatitis C in hemophiliacs were attributed to virus-inactivated factor (F) VIII concentrates despite negative results in chimpanzee experiments with regard to the transmission of hepatitis. Thus, the chimpanzee as an animal model for transmission of viral diseases was discredited as an inadequate model and even referred to as useless [1–3]. Frequently results of animal experiments are too quickly discarded for not being transferable to humans without any critical investigation into the clinical studies as such. In our view, the chimpanzee experiment yielded correct results in at least 2 of the 3 trials. It has been clarified that the batch used in one of these chimpanzee experiments was not identical with that used in the clinical trial [4]. It is very likely that in the other clinical trial transmission occurred either horizontally or by nosocomial infection [5]. Nothing is published about the third chimpanzee trial. There are only some data available from a brochure edited by the company producing the incriminated FVIII concentrate [2]. It seems that the intervals of liver biopsies are too wide, and challenge was not performed. Therefore, good compliance with the Good Laboratory Practice for preclinical tests applies to chimpanzee experiments as well. At present, the only proposed standard procedure published is for testing plasma-derived hepatitis B vaccine in the chimpanzee [6]. Already in 1987, we proposed a standard procedure for the performance of safety tests on coagulation concentrates in chimpanzees, dealing mainly with hepatitis B (HBV) and hepatitis C virus (HCV). On the basis of one chimpanzee experiment in connection with a clinical trial the problem of Good Laboratory Practice will be discussed.

Aim of the Study

A multicenter study for the safety testing of coagulation factor concentrates performed in Italy in 1984/1986 [3] demonstrated that 3 patients having been administered a vapor-treated FVIII preparation (Kryobulin® lot No. 09A058501-T) developed hepatitis B markers during the observation period of 6 months. Two patients contracted hepatitis B (alanine aminotransferase, ALT, 517 and 840 U/l, respectively), and 1 patient showed a seroconversion to hepatitis B surface (HBs) antibody only. Subtyping was successful in 1 case, revealing HBV type *ad*, which is prevalent in the USA [7, 8]. In order to prove or exclude transmission of hepatitis B by the FVIII concentrate, the same lot used for the treatment of these 3 patients was injected intravenously into thc chimpanzee. Considering the fact that these animals are endangered [9], the experiment was performed in one chimpanzee only.

Selection of the Chimpanzee

The chimpanzee chosen for this experiment was a male born in captivity (April 9, 1982). At the age of 22 months, the animal was transferred to our primate center and designated as chimpanzee E3 (ISIS No. 27). There is no difference between wild-caught and captivity-bred, or male or female chimpanzees in terms of sucseptibility to hepatitis B and hepatitis C.

Biochemical and serological markers for hepatitis B and hepatitis C of this chimpanzee were studied routinely in monthly intervals. Since October 1985, regular liver biopsies have been performed showing light-microscopically normal liver tissue and, occasionally, minimal, nonspecific reactive hepatitis which was still in compliance with our criteria. At the Hans Popper Primate Center, the chimpanzee has always been kept in an individual animal room at biosafety level 3 [10]. In the prephase, biochemical, serological and clinical findings have always been normal.

Material and Methods

Determination of Biochemical Markers

ALT (EC No. 2.6.1.2) in Serum. Healthy chimpanzees have an ALT value in serum corresponding to that of humans. Therefore, the human normal range may be taken for these tests. ALT was determined using commercially available reagent kits (Boehringer Mannheim; ALT optimized) at a reaction temperature of 25 °C (recommendation of the German Society for Clinical Chemistry [11]). For internal quality control 3 commercially available control sera were used (Monitrol I and II, Merz and Dade; Precinorm U, Boehringer Mannheim). The controls were tested at the beginning and the end of each test series.

The manufacturer defines the normal range for humans to be ≤17 U/l (women) and ≤21 U/l (men) [12, 13]. Since ALT values had always been within the normal range in the prephase, there was no reason to establish a mean +2 SD range as normal value specific for this chimpanzee. A distinct elevation of ALT in hepatitis B is reported to occur in 70–100% of experimentally infected chimpanzees [14]. An elevation of ALT about 2.5 times the upper normal limit in at least 2 consecutively drawn samples may be indicative of non-A/non-B hepatitis, provided it is confirmed by the histological findings in the corresponding liver biopsy [15].

γ-Glutamyltransferase (EC No. 2.3.2.2) in Serum. γ-Glutamyltransferase (γ-GT) is mainly present in the membrane of cells with a high secretory or absorptive capacity (kidney, liver, pancreas, intestinal brush border cells). The biological function of γ-GT is unclear. It is possible that the enzyme might be involved in the transport of amino acids or peptides through the external membrane into the cells by the γ-glutamyl cycle, as described by Meister et al. [16] in 1973. Determination of γ-GT may be used as an indication of the disturbance of excretory liver function and also had been used as marker for alcoholism, whereby minimal liver injury is demonstrable without specific histological alterations [17–19]. Since alcoholism can be definitely excluded in the chimpanzee, γ-GT is an additional marker for minimal pathological changes in the liver parenchyma [20]. In an unpublished study we observed that in 16 inoculations in chimpanzees, nearly one third of chimpanzees had no ALT elevation and about 25% had no γ-GT elevation above 2.5 times the upper normal limit, even though hepatitis C had been demonstrated by light and electron microscopy.

Thus, γ-GT was a good, additional indicator, in particular in the case of non-A/non-B hepatitis when serological test methods for detection of hepatitis C antibodies were not available. γ-GT was determined using commercially available kits of Boehringer Mannheim (Monotest Gamma-GT new [21]). Reaction temperature: 25 °C. The manufacturer defines the normal range for women as between 4 and 18 U/l, for men as between 6 and 28 U/l [22]. Several studies not yet published have shown that a normal range corresponding to the normal ALT range (upper normal limit 20 U/l) may be applicable to the chimpanzee.

Determination of Serological Markers

Hepatitis B. Serum samples were tested in codified form for HBs antigen and HBs antibody at the Max von Pettenkofer Institute in Munich (head at testing time the late Prof. E. Deinhardt). HBs-antigen-positive samples were subtyped in the same laboratory. Hepatitis B core (HBc) antibody (IgM and IgG), hepatitis B e-antigen (HBe) and HBe antibody tests were performed in our laboratory using commercially available reagent kits (Corzyme and HBe-EIA; Abbott, USA).

Hepatitis C. Serum samples stored at –20 °C were tested for HCV antibody when test reagents became available using 2nd-generation ELISA reagents from Ortho and Abbott (USA). All positive samples were additionally tested with RIBA II (immunoblot assay for the detection of antibody HCV). Results were read by eye and interpreted according to the recommendation of the producer of the reagents (Chiron Corp., USA).

Tissue Markers

Liver biopsies, using a Jamshidi needle, were performed under ketamine hydrochloride anesthesia (Ketavet®, Parke Davis; 11.5 mg/kg body weight). The material thus obtained was immediately fixed in a buffered formaldehyde solution, pH 7.0, for light microscopy, and in glutaraldehyde 3% for electron microscopy. Further processing of liver tissue is described elsewhere [23].

Samples were evaluated in codified form at the Institute of Pathology, University of Basel, in 1985–1987 by Prof. A. Bianchi and Prof. F. Gudat.

Immunohistological detection of HBs antigen in liver tissue was performed in stored formalin-fixed paraffin-embedded samples. Liver tissue was deparaffinized and afterwards equilibrated in phosphate-buffered saline buffer. HBs antigen was detected by goat anti-HBs antigen as primary antibody (Signet Laboratories Inc., USA). A universal immunoperoxidase staining kit was used as advised by the manufacturer, namely rabbit antigoat immunoglobulin as linking reagent, goat antiperoxidase-peroxidase complex and 3-amino-9-ethylcarbazide as chromogen.

Polymerase Chain Reaction

Detection of HBV DNA. A DNA sequence in the s region of HBV was detected using a standard procedure and primers consisting of 22 and 20 bases (MD 03 and MD 06), respectively. The amplified product (110 bp) was detected by a 32P end-labeled probe (MD 09) [24]. The sensitivity limit is less than 10 fg corresponding to 100 copies HBV. Several liver biopsy samples could be retrieved when polymerase chain reaction (PCR) became available. The same primers and about 4 μg liver tissue were used by test.

Detection of HCV RNA. HCV RNA is transcribed to cDNA and then amplified by PCR using nested primers of the 5′ untranslated region of the HCV genome. The amplified product (237 bp) is separated by gel electrophoresis and visualized by ethidium bromide.

Inoculum Material

On January 16, 1986, the chimpanzee E3 received a first infusion of Kryobulin lot 09A058501-T 100 U/kg body weight i.v., having a body weight of 17.66 kg at that time. Such an infusion was repeated 4 times in weekly intervals. Thus, the animal was treated with a total dose of 8,830 units FVIII. This dose is considerably above the amount administered to the 3 patients in the multicenter study for the safety testing of coagulation factor concentrates in Italy, who developed hepatitis B or seroconverted during the 6-month observation of the study. One patient received 4 infusions with a total dosage of 40 U/kg body weight (single dose 7 and 13 U/kg body weight, respectively), the second patient received 5 infusions (4 times 30 U/kg and once 60 U/kg) before detection of HBs antigen in a serum sample, and the third patient was treated with 2 infusions of 50 U/kg body weight each.

Challenge Material

An in-house HBV-positive inoculum designated as 07/85, which had already been titered in chimpanzees (10^5 chimpanzee infectious dose $(CID)_{50}$/ml), was taken as inoculum. This material was prepared from plasma of 2 patients suffering from an acute hepatitis B and contains HBe antigen subtype *ad*, HBc antibody and HBe antigen. HBV is detectable by PCR in this material using primers in the s region up to 2×10^6/ml. The very same subtype was found in 1 patient of the multicenter study, who developed hepatitis B.

Blood Sampling

Blood sampling was performed under sterile conditions from a superficial cubital vein after putting on an arm cuff to obtain compression between systolic and diastolic pressure. Serum was separated by centrifugation, and biochemical as well as serological tests were performed on the same day. Aliquots were stored at −20 °C and used for confirmation tests, determination of HCV antibody and PCR when these tests became available.

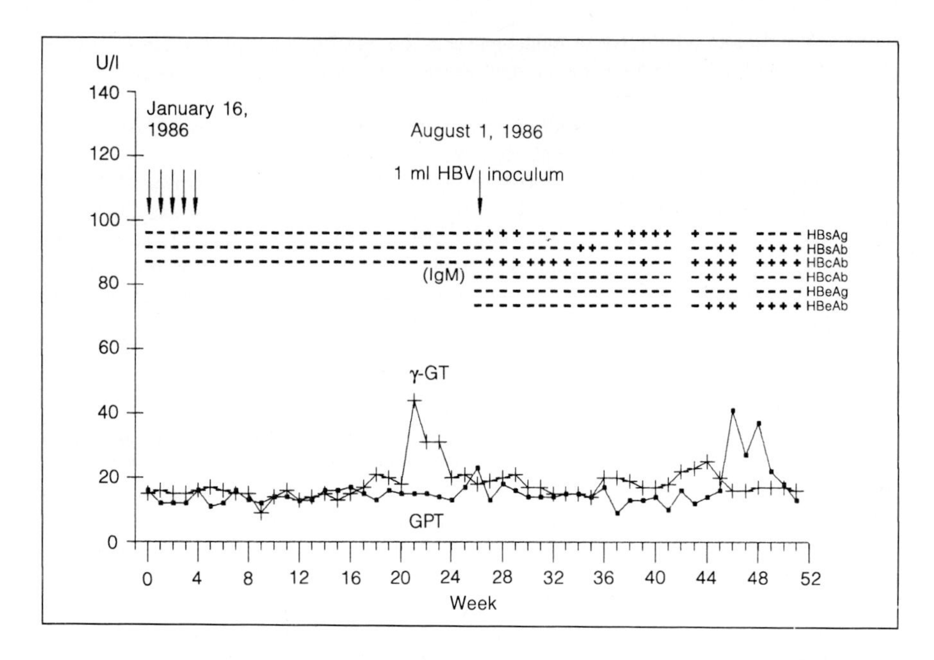

Fig. 1. Course of the transmission study in chimpanzee E3 with Kryobulin [total dose 8,830 units, 5 infusions at 100 U/kg, and challenge with 1 ml serum (10^5 CID/ml) intravenously]. GPT = Glutamic pyruvic transaminase (ALT); Ag = antigen; Ab = antibody.

Results

Treatment Phase

Over the whole 6-month observation period from January 16, 1986, until July 15, 1986, neither serological hepatitis B marker (HBs antigen, HBs antibody, HBc antibody) nor HBV DNA were detectable by PCR. ALT values were clearly in the <20 U/l range (fig. 1). In week 21 after treatment with FVIII concentrate γ-GT was 44 U/l, a value normalizing over the next weeks. This γ-GT elevation is not indicative of hepatitis C. Liver biopsies showed either normal liver tissue or a minimal nonspecific reactive hepatitis over the whole observation period. HBs antigen was not detectable in the liver tissue by immunohistochemistry (table 1).

When test methods for detection of HCV antibody became available, stored serum samples were tested. 11 samples could be retrieved from the period before the treatment with the incriminated FVIII concentrate and 77 samples during inoculation, challenge and postchallenge period. Four samples were positive in both ELISA systems used in our laboratory. However, these 4 samples were negative in RIBA II and have to be considered as false-positive results.

Table 1. Results of immunohistochemical detection of HBs antigen in hepatocytes in comparison with HBV DNA (PCR)

Biopsy date	LM diagnosis	HBs antigen in hepatocytes	HBV PCR result	
			liver	serum
Nov. 15, 1985	normal	negative	–	–
Jan. 28, 1986	NRH	negative	–	–
Mar. 5, 1986	normal	negative	–	–
Apr. 7, 1986		n.t.	–	–
Apr. 15, 1986	NRH	negative	–	–
Challenge				
Aug. 8, 1986	NRH	n.a.	n.a.	+
Aug. 13, 1986	NRH+	n.a.	n.a.	+
Aug. 22, 1986	NRH	n.a.	n.a.	+
Nov. 5, 1986	NRH min.	negative	+	+
Nov. 13, 1986	NRH	positive 1/100	+	+
Nov. 18, 1986	NRH	positive 4/100	+	+
Nov. 25, 1986	NRH	positive 4/100	+	+
Dec. 12, 1986	NRH	positive 2/100	+	+
Jan. 10, 1987	normal	negative	–	–

LM = Light microscopy; HBs-antigen-positive cells per 100 hepatocytes (immunohistochemical method as described in Material and Methods); NRH = nonspecific reactive hepatitis; n.t. = not tested; n.a. = not available.

Challenge Phase

The challenge with the hepatitis B in-house inoculum was performed on August 1, 1986, 7.5 months after the first treatment, in order to demonstrate susceptibility for hepatitis B of the chimpanzee. After inoculation of 1 ml i.v. HBs antigen was demonstrable over a period of 3 weeks, HBc antibody over 8 weeks. Quantitation of HBs antigen in these blood samples demonstrated a half-life of 5.5 days in this animal. In chimpanzees with HBs antibodies raised by vaccination or subsequent to seroconversion in the course of a B viral hepatitis, HBs antigen after intravenous injection of hepatitis-B-positive inoculum could never be demonstrated. Eleven weeks after the challenge the chimpanzee E3 developed HBs antigenemia persisting for 6 weeks. The subtype was identical with the challenge material. HBc antibody of type IgM became positive in week 18, being still detectable until week 20, which is indicative of HBV replication. HBc antibody type IgG became detectable in week 17, HBe antibody in week 18 and HBs antibody IgG in week 19 after the challenge. Examination of weekly liver biopsies showed a nonspecific reactive hepatitis only. ALT was slightly elevated on December 16, 1986, i.e. 20 weeks after the challenge, γ-GT was within the normal range over the whole period. All serum samples tested for HCV antibody were negative with one exception. The sample taken 59 weeks after inoculation and 29 weeks after challenge was positive. In RIBA II this sample was clearly negative. Several liver tissue samples were tested for HBV by PCR (table 1). All samples during the treatment phase were negative. One week after challenge up to December 12, 1986 (18 weeks after challenge), all liver samples were positive. HBs antigen could be demonstrated also in hepatocytes in a very discrete number of cells (table 1).

Discussion

It has been clearly demonstrated that neither hepatitis B nor hepatitis C was transmitted to the chimpanzee E3 by the vapor-treated FVIII concentrate. Susceptibility of the chimpanzee was confirmed by a HBV infection of identical subtype induced by challenging the animal with a low-dose HBV inoculum (10^5 CID_{50}).

Normal ALT values and normal liver histology do not indicate that susceptibility of the animal to hepatitis B was inadequate. ALT values are elevated in 70–100% of experimentally infected chimpanzees using high-dose inocula [14]. The percentage of histologically detected acute hepatitis B in weekly performed liver biopsies may vary due to different inocula. As few as 33% up to 82% of acute hepatitis B have been reported by several investigators [25, 26]. There is no relation between the concentration of the inoculum and the histological findings. The only reliable markers for hepatitis B infection of the chimpanzee are the serological markers occurring in 100% of experimentally infected chimpanzees [25]. We therefore consider HBs antigenemia followed by the occurrence of HBc antibody sufficient for demonstrating susceptibility of the chimpanzee E3. During the whole observation period of the treatment phase (6 months) and the challenge phase neither HCV antibodies nor HCV RNA could be demonstrated by PCR. Therefore the vapor-heated FVIII concentrate in the dosage used does not transmit hepatitis B or hepatitis C. Mannucci et al. [27] reported on HCV infection in a clinical trial which resulted in hepatitis B infection in 3 patients with the same FVIII concentrate, which clearly did not transmit hepatitis B or C in chimpanzees. The diagnosis was based on HCV antibody detection by ELISA and RIBA II. Three other patients – negative for hepatitis B – were found positive for hepatitis C. The authors report that when these patients were interviewed again, it turned out that at least two of them had definitely been infused with unspecified amounts of blood products before entering the study. One of the former hepatitis-B-positive patients had been found to develop HCV antibodies in week 12 after treatment, being negative at entry into the study. It is concluded that this is the first reported double infection in a hemophiliac. A double infection caused by the inoculum should have been noticed in the chimpanzee experiment. Simultaneous infections with hepatitis B and C have resulted only in a difference in the incubation period of a few weeks [25]. Interference between hepatitis B and C in the chimpanzee model has only been discussed if chronic HBV carrier chimpanzees are infected with hepatitis C or vice versa [28–30]. Several hepatitis C episodes in a series of chimpanzee experiments have been reported, demonstrating that this animal model is highly sensitive for HCV [31]. Prince et al. [32] and Farci et al. [33] confirmed these data using PCR for detection of HCV RNA. Moreover the specificity of the HCV antibody tests is disputed. There is some evidence of false-positive test results in humans [34]. However, most of the questionable positive results are in connection with autoimmune hepatitis [35]. At least we should not forget that the source material of all the HCV test methods was gained from a chimpanzcc [36].

Conclusion

The study described here demonstrates clearly the importance of good laboratory practice for chimpanzee experiments as well as for human trials. It is critical to use 'naive' individuals for all transmission experiments.

In order to guarantee proper safety testing the following prerequisites are inevitable: absolute healthy conditions of the chimpanzees used in the experiment; other experiments must not be simultaneously performed in the same animal; regular health check for parasites, tuberculosis, calcium metabolism and hematologic disorders is required; the chimpanzees must be housed without any exception in individual cages in laboratories of biosafety level 3; particular attention must be paid to avoid nosocomial infections; employment of highly qualified animal keepers, cleaning and kitchen personnel is essential. It is the state of the art that the employees are immunized against hepatitis B and that regular health checks are carried out, including X-ray, liver function test, urine test and feces test for parasites. Blood sampling and liver biopsies in chimpanzees have to be performed under strict sterile conditions. Any risk of cross-contamination caused by blood must be ruled out. Inoculation must be performed in such a way that clinical treatment of patients is simulated as closely as possible. Any chimpanzee experiment is invalid if the susceptibility of the chimpanzee is not tested with a titrated challenge inoculum. It would be a real achievement, if standard inocula were available as international reference preparation for challenge experiments in chimpanzees.

References

1 Colombo M, Mannucci PM, Carnelli V, Savidge GF, Gazengel C, Schimpf K, the European study group: Transmission of non-A, non-B hepatitis by heat-treated factor VIII concentrate. Lancet 1985;ii:1–4.
2 Preston FE, Hay CRM, Dewar MS, Greaves M, Triger DR: Non-A, non-B hepatitis and heat-treated factor VIII concentrates. Lancet 1985;ii:213.
3 Mannucci PM, Zanetti AR, Colombo M, the Study Group of the Fondazione dell'Emophilia: Prospective study of hepatitis after factor VIII concentrate exposed to hot vapour. Br J Haematol 1988;68:427–430.
4 Hollinger FB, Dolana G, Thomas W, Gyorkey F: Reduction in risk of hepatitis transmission by heat treatment of a human factor VIII concentrate. J Infect Dis 1984;150:250–262.
5 Eibl J, Eder G, Anderle K: Issues associated with safety testing of blood products in endemic areas. Br J Haematol 1988;69:575–576.
6 Berthelot P, Courouce AM, Eyquem A, Feldman G, Jacob J, Ravisse P, Vacher B, Moor-Jankowski J, Muchmore E, Prince A: Hepatitis B vaccine safety monitoring in the chimpanzee: Interpretation of results. J Med Primatol 1984;13:119–133.
7 Mazzur S, Burgert S, Blumberg BS: Geographic distribution of Australian antigen determinants *d*, *y* and *w*. Nature 1974;247:38–40.
8 Holland PV: Hepatitis B surface antigen and antibody (HBs Ag/anti HBs); in Gerety RJ (ed): Hepatitis B. New York, Academic Press, 1985, pp 15–17.
9 Convention on International Trade in Endangered Species of Wild Fauna and Flora, signed in Washington, March 3, 1973.
10 Richardson JH, Barkley WE (eds): Biosafety in Microbiological and Biomedical Laboratories, ed 2. HHS Publ No (CDC) 84-8395. US Department of Health and Human Services, Public Health Service, Center for Disease Control and National Institute of Health, 1988.
11 Empfehlungen der Deutschen Gesellschaft für klinische Chemie. Z Klin Chem klin Biochem 1972; 10:182–192.
12 Wallnoefer H, Schmidt E, Schmidt FW (eds): Synopsis der Leberkrankheiten. Stuttgart, Thieme, 1974.
13 Thefeld W, Hoffmeister H, Busch EW, Koller PU, Vollmar J: Referenzwerte für die Bestimmung der Transaminasen GOT und GPT sowie der alkalischen Phosphatase im Serum mit optimierten Standardmethoden. Dtsch Med Wochenschr 1974;99:343–351.
14 Tabor E, Purcell RH, London WT, Gerety RJ: Use of an interpretation of results using inocula of hepatitis B virus with known infectivity titers. J Infect Dis 1983;147:531–534.
15 Mannucci PM, Colombo M, Bianchi A: Revision of the protocol recommended for studies of safety from hepatitis of clotting factor concentrates. Thromb Haemost 1989;61:532–534.
16 Meister A, Tate SS, Ross LL: Membrane-bound gammaglutamyl-trans-peptidase; in Martonosi A (ed): The Enzymes of Biological Membranes. New York, Plenum Press, 1976, vol III, pp 315–347.
17 Rosalki SB: Gamma-Glutamyltranspeptidase. Adv Clin Chem 1975;7:53–107.
18 Rosalki SB, Rau D: Serum gamma-glutamyltranspeptidase activity in alcoholism. Clin Chim Acta 1972;39:41–47.
19 Weill J, Schellenberg F, LeGoff AM, Lamy J: The predictive value of gamma-glutamyltransferase and other peripheral markers in the screening of alcohol abuse; in Siest G, Heusghem C (eds): Gammaglutamyltransferases: Advances in Biomedical Pharmacology, 3rd ser. Paris, Masson, 1982, pp 195–198.
20 Valenza FP, Muchmore E: The clinical chemistry of chimpanzees. II. Gamma glutamyl transferase levels in hepatitis studies. J Med Primatol 1985;14:305–315.
21 Persijn JP, van der Slik W: A new method for the determination of gamma-glutamyltransferase in serum. J Clin Chem Clin Biochem 1976;14:421–427.
22 Wuest H: Laboruntersuchungen; in Demling L (ed): Klinische Gastroenterologie in 2 Bänden, ed 2. Stuttgart, Thieme, 1984, vol I: Diagnostische Übersicht, Mundhöhle und Rachen, Speiseröhre, Magen, Darm.
23 Gudat F, Bianchi L, Sonnabend W, Thiel G, Aenishaenslin W, Stalder GA: Pattern of core and surface expression in liver tissue reflects a state of specific immune response in hepatitis B. Lab Invest 1975;32:1–9.

24 Baginski I, Ferrie A, Watson R, Mack D: Detection of hepatitis B virus; in Innis MA, Gelfand DH, Sninsky JJ, White TJ (eds): PCR Protocols: A Guide to Methods and Application. San Diego, Academic Press, 1990, p 348.
25 Tabor E: Development and application of the chimpanzee animal model for human non-A, non-B hepatitis; in Gerety RJ (ed): Non-A, Non-B Hepatitis. New York, Academic Press, 1981, pp 189–206.
26 Tabor E, Purcell RH, Gerety RJ: Primate animal models and titered inocula for the study of human hepatitis A, hepatitis B, and non-A, non-B hepatitis. J Med Primatol 1983;12:305–318.
27 Mannucci PM, Zanetti AR, Colombo M, Chistolini A, De Biasi R, Musso R, Tamoni G, for the Study Group of the Fondazione dell'Emofilia: Antibody to hepatitis C virus after a vapour-heated factor VIII concentrate. Thromb Haemost 1990;64:232–234.
28 Brotman B, Prince AM, Huima T, Richardson L, van den Ende MC, Pfeifer U: Interference between non-A, non-B and hepatitis B virus infection in chimpanzees. J Med Virol 1983;11:191–205.
29 Tsiquaye KN, Portman B, Tovey G, Kessler H, Shanlian H, Xioa-Zhen L, Zuckerman AJ, Craske J, Williams R: Non-A, non-B hepatitis in persistent carriers of hepatitis B virus. J Med Virol 1983; 11:179–189.
30 Bradley DW, Maynard JE, McCaustland KA, Murphy BL, Cook EH, Ebert JW: Non-A, non-B hepatitis in chimpanzees: Inteference with acute hepatitis A virus and chronic hepatitis B virus infection. J Med Virol 1983;11:207–213.
31 Eder G, Bianchi L, Gudat F: Transmission of non-A, non-B hepatitis to chimpanzees: A second and third episode caused by the same inoculum; in Zuckerman AJ (ed): Vir Hepatitis Liver Dis. Liss Inc., New York, 1988, pp 550–552.
32 Prince AM, Brotman B, Huima T, Pascual D, Jaffery M, Inchauspé G: Immunity to Hepatitis C Infection. New York, J Inf Dis 1992;165:438–443.
33 Farci P, Alter HJ, Ogata N, Wong D, Engle R, Miller R, Dawson G, Lesmiewski R, Mushahwar I, Purcell R: Lack of protection against reinfection with hepatitis C virus (HCV) in multiple cross-challenges of chimpanzees. Lecture at the Postgraduate Course and 42nd Annual Meeting of the American Association for the Study of Liver Diseases. Hepatology 1991;14(pt 2):90A.
34 Gray JJ, Wreghitt TG, Friend PJ, Wight DGD, Sundaresan V, Calne RY: Differentiation between specific and non-specific hepatitis C antibodies in chronic liver disease. Lancet 1990;335:609–610.
35 Schvarcz R, Von Sydow M, Weiland O: Autoimmune chronic active hepatitis: Changing reactivity for antibodies to hepatitis C virus after immunosuppressive treatment. Scand J Gastroenterol 1990;25:1175–1180.
36 Choo QL, Kuo G, Weiner AJ, Overby LR, Bradley DW, Houghton M: Isolation of a cDNA clone derived from a blood-borne non-A, non-B viral hepatitis genome. Science 1989;244:359–362.

Gerald Eder, MD, Hans Popper Primate Center, Head, Clinical Research-Gastroenterology, Immuno AG, Uferstrasse 15, A–2304 Orth/Donau (Austria)

Eder G, Kaiser E, King FA (eds): The Role of the Chimpanzee in Research.
Symp, Vienna 1992. Basel, Karger, 1994, pp 166–168

The Role of Chimpanzees in Research on Slow Infections of the Human Nervous System

The Spongiform Encephalopathies

David M. Asher, Clarence J. Gibbs Jr., Alfred E. Bacote, Michael P. Sulima, D. Carleton Gajdusek

Laboratory of Central Nervous System Studies, Division of Intramural Research, National Institute of Neurological Disorders and Stroke, National Institutes of Health, Bethesda, Md., USA

Chimpanzees played an important role in the original demonstration that the spongiform encephalopathies of humans [1] were slow infections [2]. Brain suspensions from 11 patients with kuru transmitted progressive encephalopathy [3] to 20 chimpanzees, as did suspensions from 34 patients with Creutzfeldt-Jakob disease [4] to 45 chimpanzees and that of a patient with Gerstmann-Sträussler syndrome [5] to 1 chimpanzee. The striking similarities of clinical illnesses and histopathological changes in the chimpanzees to those of the patients helped to convince the medical community to accept the infectious etiology of diseases once considered to be idiopathic degenerative conditions. Chimpanzees were important in showing that the infectious agents of spongiform encephalopathies were present in very large amounts in brains of patients (more than 300 million lethal doses per gram of tissue) and in some other tissues [6] in smaller amounts. Chimpanzees were initially infected by intracerebral inoculation of human tissue suspensions and later by less direct routes of inoculation as well. Chimpanzees were used to demonstrate that the Kuru agent is filterable and serially transmissible from one animal to another without reduction in titers of infectivity – confirming that it was a replicating agent smaller than bacteria. Important properties of the human agents were also first established by inoculation into chimpanzees – resistance to formaldehyde and to neutralization by patients' sera. The potential danger of exposure to the agent of Creutzfeldt-Jakob disease by medical personnel and patients was first confirmed by studies in which chimpanzees were infected by inoculations of brain suspensions from a neurosurgeon [7] and several patients who had undergone stereotactic surgery or therapy with human pituitary growth hormone [8]. (Stereotactic electrodes incriminated in those cases also transmitted disease to a chimpanzee.) Chimpanzees were also used to confirm that in brain tissues of patients with familial Creutzfeldt-Jakob disease expressed in a pattern typical of an autosomal dominant genetic disorder [9] a transmissible agent was also present. From the earliest years of study of the human spongiform encephalopathies, attempts were made to find better assays for the infectious agents, by inoculations of cell cultures and various animals. (A chimpanzee was used to confirm the first successful transmission of kuru to a monkey.) Cell cultures have been of limited use in studying the spongiform encephalopathies. However, several species of animals, especially squirrel monkeys (for primary transmission of the agent from human material), mice and hamsters (to characterize properties of the agents using selected rodent-adapted strains), have almost completely replaced chimpanzees for studying the spongiform encephalopathies. The use of chimpanzees for studies of slow infections of the human nervous system remains justified in unusual circumstances, for example, seeking infectious agents in diseases of unknown etiology; however, none of several common neurological diseases of humans – multiple sclerosis, amyotrophic lateral sclerosis, Alzheimer's disease and others – has been successfully transmitted to chimpanzees [10] or other animals in confirmed attempts.

References

1 Gajdusek DC: Subacute spongiform encephalopathies: Transmissible cerebral amyloidoses caused by unconventional viruses; in Fields BN, Knipe DM (eds): Virology, ed 2. New York, Raven Press, 1990, pp 2289–2324.
2 Sigurdsson B: Observations on three slow infections of sheep. Br Med J 1954;110:255–270, 307–322, 341–354.
3 Gajdusek DC, Gibbs CJ Jr, Alpers M: Experimental transmission of a kuru-like syndrome in chimpanzees. Nature 1966;209:794–796.
4 Gibbs CJ Jr, Gajdusek DC, Asher DM, et al: Creutzfeldt-Jakob disease (spongiform encephalopathy): Transmission to the chimpanzee. Science 1968;161:388–389.
5 Masters CL, Gajdusek DC, Gibbs CJ Jr: Creutzfeldt-Jakob disease virus isolation from the Gerstmann-Sträussler syndrome, with an analysis of the various forms of amyloid deposition in the virus-induced spongiform encephalopathies. Brain 1981;104:559–588.
6 Asher DM, Gibbs CJ Jr, Gajdusek DC: Pathogenesis of spongiform encephalopathies. Ann Clin Sci 1976;6:84–103.
7 Gajdusek DC, Gibbs CJ Jr, Earle K, Dammin GJ, Schoene WC, Tyler HR: Transmission of subacute spongiform encephalopathy to the chimpanzee and squirrel monkey from a patient with papulosis maligna of Köhlmeyer-Degos. Excerpta Med Int Congr Ser 1974;319:390–392.

8 Brown P: Iatrogenic Creutzfeldt-Jakob disease. Aust NZ J Med 1990;20:633–635.
9 Asher DM, Masters CL, Gajdusek DC, Gibbs CJ Jr: Familial spongiform encephalopathies; in Kety S, Rowland L, Sidman R, Matthysse S (eds): Genetics of Neurological and Psychiatric Disorders. New York, Raven Press, 1983, pp 273–291.
10 Goudsmit J, Morrow CH, Asher DM, et al: Evidence for and against the transmissibility of Alzheimer's disease. Neurology 1980;30:945–950.

David Asher, MD, Laboratory of Central Nervous System Studies, Basic Neurosciences Program, Division of Intramural Research, National Institute of Neurological Disorders and Stroke, Bldg 35, Rm 5B-21, National Institutes of Health, Bethesda, MD 20893 (USA)

Eder G, Kaiser E, King FA (eds): The Role of the Chimpanzee in Research.
Symp, Vienna 1992. Basel, Karger, 1994, pp 169–170

Retrolentivirus Infections in Man and Chimpanzees

Close Enough to Be Different

P.L. Nara

Virus Biology Unit, LTCB, NCI-FCRDC, Frederick, Md., USA

Natural infection with retrolentiviruses results generally in a family-genus-species-restricted, chronic-active, lifelong infection of the lymphoreticulo-endothelial system. Despite this family-genus-restricted behavior, members of the lentiviral subfamily are found to infect a wide array of vertebrates ranging phylogenetically from ungulates to carnivores to nonhuman and human primates. This chronic-active, persistent viral-host interaction appears to exist as both a host adapted (apathogenic) and host-unadapted (pathogenic) state depending on as yet undetermined 'pathogenic factors'. These pathogenic factors appear to include a complicated array of both host and viral determinants. In host-adapted species the lack of disease does not appear to directly correlate with any one or simple combinations of the following: the degree of antigenic variation, the nature and strength of the antiviral immune response and/or cell types infected. The chimpanzee *(Pan troglodytes)* is the closest living relative of modern man and is readily infected in vivo by various sources of human tissue and viral HIV-1-containing material. Despite its exquisite sensitivity to viral infection, and the apparent similarity of the density of the CD4 receptor on peripheral T4 cells, no sustained immunopathologic clinical sequelae have developed 8 years after experimental infection. Lack of disease to date appears to involve a multifactorial process possibly involving both host and viral factors. These factors include: compartmentalization of infection, low viral load in circulating and lymphoid tissues, absence of plasma viremia, CD8 cells capable of suppressing viral replication, limited in vitro viral replication and spread, minimal to absent cytopathology of infected T4 cells in the presence of other peripheral blood cell populations, and a single amino acid position difference in the CD4 molecule responsible for syncytium formation but not infectivity, the absence of gp120-induced apoptosis, the absence of anti-CD4 antibody or other host-specific cellular antibody suggesting the induction of autoimmunity, the lack of evidence for infecting critical stem cells throughout the lymphoid and blood-forming organs and the presence of a vigorous immune response to the virus. Thus host adaptation with retrolentiviruses in primates may represent a balance between the virus load in critical end-organs in the body, the breaking tolerance due to molecular mimicry and/or the slight alteration of 'self' within the context of the MHC repertoire of the immune system and/or minor amino acid differences in cellular receptor molecules such as CD4.

Regarding the ancestral relationships and origin of HIV-1, the simian immunodeficiency virus type 1 chimpanzee ($SIV\text{-}1_{cpz}$) represents the closest naturally occurring isolate of a primate virus to date with 84% overall homology. In its natural host SIV_{cpz} appears also to exist as a host-adapted species; however, only 2 chimpanzees out of 83 tested in the Gabon colony have been found seropositive and virus isolation positive and as such represent a much too limited data set. It may represent still an intermediate ancestor of HIV-1 with other more closely related isolates existing in the other 6 known wild subspecies of chimpanzees or other members of the great ape family.

The experimental model of HIV-1 infections in chimpanzees has provided the field with critical and important insights into the kinetics of the humoral response, virus neutralization, vaccine immunogenicity studies, functional epitope mapping, conformational antibodies, a model for clonal dominance, virus selection, virus variation and viral elimination. The further understanding of the natural ecology, biology and behavior both of the wild and colony captive-protected animals should allow for important future biomedical investigations to be made which in the most pessimistic of settings should improve the health of both man and man's closest living relative.

Peter L. Nara, DVM, PhD, Acting Head, Virus Biology Unit, LTCB, NCI-FCRDC, Bldg 560, Rm 12-42, Frederick, MD 21702–1201 (USA)

Eder G, Kaiser E, King FA (eds): The Role of the Chimpanzee in Research.
Symp, Vienna 1992. Basel, Karger, 1994, pp 171–174

Studies on Prophylaxis against Hepatitis B Using the Chimpanzee Model

Sten Iwarson

Department of Infectious Diseases, University of Göteborg, Sweden

Introduction

The chimpanzee model has been used for several research purposes within the field of viral hepatitis. Development of safe and effective hepatitis B vaccines and immunoglobulins (containing high titers of antibodies against the disease) became possible in the 1970s using chimpanzees as experimental animals.

Hepatitis B, an inflammation of the liver caused by hepatitis B virus (HBV), is characterized by the appearance of markers of infection (antigens and antibodies) in the serum of humans and chimpanzees. Usually, chimpanzees do not have clinically recognized symptoms or signs of the disease, but elevated levels of aminotransferases in serum (aspartate aminotransferase and alanine aminotransferase, ALT) as a sign of damaged liver cells are seen in chimpanzees as well as in humans.

When chimpanzees are used for studies of safety or efficacy of vaccines as in the present one, the animals must be fully susceptible to HBV. Ideally all chimpanzees should be born and raised in a breeding colony with continuous testing of serological hepatitis B markers [1].

Material and Methods

Chimpanzees

All chimpanzees used in these studies were born in a breeding colony in the USA (Primate Research Institute, New Mexico State University, Alamogordo, N. Mex., USA), under Food and Drug Administration contract No. 223-85-1004. None of the chimpanzees had previously been exposed to hepatitis B, and all had normal serum aminotransferase (ALT) activities and normal histology on liver biopsy prior to initiation of the study. The studies were carried out at the Office of Biologies (FDA), Bethesda, Md., USA.

Serological and Biochemical Analyses

Serum specimens were taken weekly from the animals beginning several weeks before and continuing for 1 year after HBV inoculation. ALT analyses were performed by a spectrophotometric method (normal level ≤ 40 IU/l). HBV markers – hepatitis B surface antigen (HBsAg), antibodies to HBsAg (anti-HBs) and hepatitis B core antigen (anti-HBc) – were performed with commercial radioimmunoassays (Ausria II, Ausab and Corab; Abbott Laboratories, USA).

Liver Biopsies

Liver biopsies were obtained approximately every 3 weeks with a Vim-Silverman needle. Tissue specimens were fixed in 10% formalin for light microscopy studies. The histological definitions of viral hepatitis used were those presented by an international group.

HBV Inoculum

The HBV inoculum used (NCDB/NIAID) has been described previously [1]; this inoculum has caused hepatitis B in 100% of approximately 50 previously exposed chimpanzee infectious doses (CID_{50}) of HBV subtype ayw. This inoculum was given intravenously to each of the 5 chimpanzees simultaneously with or following immunoprophylaxis.

Study 1: Failure of Hepatitis B Immune Globulin to Protect against Experimental Hepatitis B Infection in Chimpanzees

To study experimentally the protective effect of postexposure prophylaxis against hepatitis B [2], a special preparation of hepatitis B immune globulin (HBIG) was injected intravenously into 3 chimpanzees simultaneously with, or at different time intervals after, intravenous injection of a titered inoculum of HBV. The HBIG was given either simultaneously with the HBV inoculum, at 4 h after or at both 4 h and 4 weeks after the HBV injection. A fourth chimpanzee received a standard preparation of HBIG intramuscularly at both 4 h and 4 weeks after receiving the HBV injection. A fifth animal received HBIG intravenously 4 h after the HBV inoculum and at the same time received its first of 3 hepatitis vaccine injections. All chimpanzees were followed for 1 year.

The animals which received HBIG simultaneously with HBV or received HBIG plus vaccine had no serological or biochemical sign of hepatitis during follow-up. The 3 animals which received HBIG inoculation all developed HBs antigenemia and serum ALT elevations. HBsAg did however appear in serum several weeks later than expected for the HBV inoculum used. Postexposure prophylaxis with HBIG protected the HBV-exposed chimpanzees only if HBIG was combined with hepatitis B vaccination or if HBIG was given simultaneously with the HBV inoculum.

Study 2: Immunization with Hepatitis B Core Antigen

Although anti-HBs usually provides protection against HBV infection, recent reports indicate that this is not always the case [3]. To study the possible role of immune responses to hepatitis B core antigen (HBcAg) in immunity to HBV infection, chimpanzees were immunized with chimpanzee liver-derived or genetically cloned HBcAg and later challenged with known infectious HBV. Two chimpanzees, which received liver-derived or cloned HBcAg in Freund's adjuvant and developed anti-HBc and low-titer hepatitis B antibody, were completely protected against HBV infection following challenge. In contrast, another chimpanzee, which received liver-derived HBcAg without adjuvant, developed anti-HBc only in serum and had a subclinical HBV infection when challenged.

These findings demonstrate that protection against HBV infection can be induced by immunization with HBcAg in adjuvant and that protection, in this case, is not solely dependent on anti-HBs. This fact has important implications in our understanding of the biology of HBV infection and in the design of future hepatitis B vaccines.

Study 3: Postexposure Vaccination against Hepatitis B

To study the effect of vaccination against hepatitis B after exposure to the virus [4], 4 chimpanzees were vaccinated with Merck's plasma-derived hepatitis B vaccine 4, 8, 48 and 72 h, respectively, after intravenous injection of an infectious HBV inoculum. The second and third vaccine inoculations were given 2 and 6 weeks later, i.e. at considerably shorter intervals than recommended either for ordinary prophylactic vaccination or for postexposure vaccination in combination with HBIG.

The chimpanzees were followed for 1 year. None showed HBs antigenemia, liver enzyme elevation (ALT) or histopathological alterations in liver biopsies. Late appearance of anti-HBs was observed only in the serum of the animal whose series of vaccination started 72 h after HBV inoculation. An unvaccinated control chimpanzee, which received the HBV inoculum only, developed clinical hepatitis B with ALT elevations and HBs antigenemia within 2 months of the experimental HBV exposure.

These results seem to indicate that postexposure vaccination against hepatitis B, begun within 48 h after HBV exposure and with short intervals between the vaccine injections, can protect against hepatitis B infection also when concomitant HBIG prophylaxis is not given, at least in this experimental model.

We have used this postexposure vaccination model for 3 years in hospital personnel following accidental HBV exposure in our hospital department, and our experiences are good.

It may be argued that vaccination only does not work as well in HBV-exposed newborns as found in these studies. However, the situation of accidental exposure in adults is different from the vertical transmission situation. Some newborns probably have intrauterine HBV exposure long before birth, and vaccination only should not work in these cases. Also, HBV exposure may be heavier in newborns. So, the combination of HBIG and hepatitis B vaccination at a reasonably rapid vaccination schedule will probably protect HBV-exposed newborns in the best way [5].

Conclusions

HBV-associated antigens other than HBsAg can stimulate immunity against HBV infection, and this fact is becoming even more important as reports of failures of vaccination with the current hepatitis B vaccine (containing only HBsAg) begin to appear in the literature and as HBV infections occur in individuals with preexisting anti-HBs in serum.

The observations on postexposure prophylaxis against hepatitis B raise two important questions. Firstly, does HBIG alone – except perhaps in neonates – have any useful protective effect when given after exposure to HBV? Secondly, does accelerated postexposure vaccination alone, without concomitant HBIG prophylaxis, offer equally effective protection after low-dose HBV inoculation, as in needle-prick injuries? The possibility that an accelerated vaccination schedule may protect against hepatitis B when started within 48 h after HBV exposure requires further investigation in both adults and infants born to HBV-infected mothers.

References

1 Tabor E: Chimpanzee model for the study of hepatitis B; in Gerety RJ (ed): Hepatitis B. Orlando, Academic Press, 1985, pp 303–317.
2 Wahl M, Iwarson S, Snoy P, Gerety RJ: Failure of hepatitis B immune globulin to protect against experimental infection in chimpanzees. J Hepatol 1989;9:198–203.
3 Iwarson S, Tabor E, Thomas HC, Snoy P, Gerety RJ: Protection against hepatitis B virus infection with hepatitis B core antigen. Gastroenterology 1985;88:763–767.
4 Iwarson S, Wahl M, Ruttiman E, Snoy P, Seto B, Gerety RJ: Successful postexposure vaccination against hepatitis B in chimpanzees. J Med Virol 1988;25:433–439.
5 Iwarson S: Post-exposure prophylaxis for hepatitis B: Active or passive? Lancet 1989;ii:146–147.

Sten Iwarson, MD, PhD, Department of Infectious Diseases, University of Göteborg, Östra Hospital, S–41685 Göteborg (Sweden)

Eder G, Kaiser E, King FA (eds): The Role of the Chimpanzee in Research.
Symp, Vienna 1992. Basel, Karger, 1994, pp 175–179

Hepatitis B in Chimpanzees

Daniel Shouval

Liver Unit, Hadassah University Hospital, Jerusalem, Israel

Introduction

Hepatitis B virus (HBV) infection affects hundreds of millions of people worldwide. During the past 25 years, enormous progress has been achieved in the characterization of HBV, the pathophysiology of disease in humans and its prevention. Initial progress in research of HBV was, however, slow, mainly for two reasons: (a) the virus cannot be grown in tissue culture; (b) the lack of an adequate animal model system. Attempts to transmit hepatitis viruses to commonly used laboratory animals, including mice, guinea pigs and minipigs, already failed 30–50 years ago [1], and the initial experiment to infect non-human primates failed, as well. In fact, the first incidence of propagation of 'hepatitis' from non-human primates to humans occurred most probably in 1938, when almost 30% of individuals receiving an experimental yellow fever vaccine containing rhesus monkey serum developed jaundice [1]. The causative agent was not identified at the time, and it took another 30 years until unequivocal evidence was gathered which linked non-human primate handlers with an increased risk to contract hepatitis. Deinhardt [1] has reviewed 46 small clusters with 220 cases of viral hepatitis in individuals who were in close contact with non-human primates. Most of the outbreaks occurred in association with recently imported chimpanzees and a few in association with contact with a gorilla, Celebes apes and woolly monkeys [1]. Once the differences between hepatitis A, B and non-A, non-B viruses were recognized, it became evident that the chimpanzee is one of the few non-human primates susceptible to HBV infection [2–4]. Some of the chimpanzees involved in the early outbreaks of hepatitis in humans received so-called 'passive immunization' with pooled human plasma while still in Africa, which was intended to protect them against human diseases in the west. It is therefore unclear whether the

Table 1. HBV in experimental animals

'True' infection	Seroconversion without infection	Artificial hosts for HBV+ cells
Chimpanzee	Gibbons	Athymic mice
Gorillas	African green monkeys	Athymic rats
Orangutans		Transgenic mice
Woolly monkeys (?)		
Rhesus monkey (?)		

chain of HBV infection in chimpanzees started in the non-human primates or in humans. Later it was established that, except for anecdotal reports of HBV infections in rhesus monkeys, gorillas or orangutans, the chimpanzee is the only non-human primate which can reproducibly be infected with HBV (table 1). Occasional seroconversion to hepatitis B surface antibody (anti-HBs), but not true infection, has been documented in gibbons and green African monkeys. Artificial model systems have been described in which hepatocellular carcinoma cell lines containing integrated or episomal HBV DNA, as well as transfected HBV cells, have been transplanted into athymic mice or rats. Transgenic mice bearing HBV DNA and expressing specific viral proteins have also been developed, and small animal model systems susceptible to HEPADNA viruses were described (table 1). However, HBV infection in most of these animal models often bore little similarities with infection in humans. The chimpanzee is therefore for all practical purposes the only non-human primate susceptible to HBV infection. Therefore, despite the many difficulties in raising chimpanzees and the high cost involved, their contribution to the prevention of HBV infection in humans is prodigious.

The Clinical Course of HBV Infection in Chimpanzees

Since the first published reports on HBV infection in chimpanzees [2–6], it gradually became apparent that these non-human primates were indeed highly susceptible to HBV infection, and the success of challenges is over 90% using similar inoculation doses as compared to humans accidentally exposed. However, the infection follows a much milder course as compared to humans. Moreover, the severity of the clinical disease is virtually unaffected by the infectious dose, although the length of the incubation period can be shortened by increasing the infecting dose. Following inoculation of well-characterized human serum positive for hepatitis B surface (HBsAg) and hepatitis B e-antigen (HBeAg), first detection of HBsAg, hepatitis B core antibody (anti-HBc) and HBeAg may occur between 3 and 17 weeks after inoculation. HBsAg persists for several days up to 44 weeks, followed by seroconversion to anti-HBs, which occasionally appears weeks to months after HBsAg becomes undetectable [1, 6, 7]. Alanine aminotransferase or aspartate levels usually rise to several hundred, and occasionally to above 1,000 units approximately 2–4 months after inoculation. The histopathologic manifestations of acute viral hepatitis B in chimpanzees are similar to those in humans. Popper et al. [8] studied consecutive liver biopsies of 7 chimpanzees acutely infected with HBV. Observed changes include degeneration of hepatocytes with necrosis, lobular hepatitis and portal inflammatory activity, as well as activation of sinusoidal lining cells and detection of macrophages. The degree of lobular inflammatory activity and reaction seems to correlate with alanine aminotransferase activity. Most histological changes resolve in parallel with improvement of hepatocellular injury as manifested by alanine aminotransferase normalization; and, as observed in humans, the inflammatory infiltrates retract to the portal spaces before complete resolution.

Shikata et al. [9] suggested to separate experimental HBV infection in chimpanzees into two distinct types according to the biochemical, serologic, and histopathologic manifestations. The first type is defined by a self-limited, rapidly resolving hepatitis with spiking and short term rise in alanine aminotransferase, which begins approximately 5 weeks after detection of HBsAg. The second type is characterized by a low-grade persistent hepatitis which occurs about 10 weeks after appearance of HBsAg and has a more pronounced portal lymphocytic infiltration as compared to the first type. The risk of a chimpanzee experimentally infected with HBV to develop an HBsAg carrier state is not established. There are, however, several reports on HBsAg chimpanzee carriers in the literature, in whom abundant HBsAg with ground glass appearance and occasional hepatitis B core antigen were identified, in cytoplasm and nuclei, respectively [1, 6, 9, 10].

The natural history of HBV infection in carrier chimpanzees is characterized by normal or mildly elevated aminotransferase levels and histopathologic manifestation of chronic persistent hepatitis. We have previously examined serum and percutaneous liver biopsies from 10 chimpanzees with acute HBV infection, 5 animals with HBV carrier states documented for 4–14 years and 5 normal animals [11]. All 5 HBsAg carriers (3 females and 2 males) were anti-HBc+/HBeAg+, and 1 animal was also positive for hepatitis B envelope antibody; mild alanine aminotransferase elevation was observed in 2 of the chimpanzee carriers. In 4 of 5 carriers there was histologic evidence for chronic persistent hepatitis. Episomal HBV DNA was found in nucleic acid extract of percutaneous liver biopsies in all 5 HBsAg carriers, with no evidence for integration of the viral DNA into the host genome.

Comment

The important contribution of the chimpanzee to the research and eradication of hepatitis virus infections A, B and C is reflected in over 400 published reports during the past 15 years. Despite some major dissimilarities with HBV infection in humans, the chimpanzee model system for HBV has facilitated our knowledge as to infectivity of defined HBV inocula, immunopathognesis, resistance to challenge, clearance of infection, the molecular characterization of HBV mutants and δ virus. Chimpanzees played a major role in the development of passive prophylaxis and active vaccination against HBV. However, experiments to test new antiviral agents against HBV in chimpanzees have so far been unrewarding. It is important to note that the HBsAg carrier state in the chimpanzee is not associated with repeated cycles of massive necrosis, fibrosis and regeneration, as observed in humans. Furthermore, despite an occasional observation of an inflammatory infiltrate spillover across the lining of some portal tracts, chronic active hepatitis or cirrhosis has not been observed in chimpanzee carriers, who have been followed for more than a decade. Moreover, hepatocellular carinoma has never been reported in chronic HBV chimpanzee carriers demonstrating continuous viral replication over many years. The immunopathogenesis of the carrier state in HBsAg+ chimpanzees is poorly understood, and the cellular as well as humoral defects described in human HBsAg carriers have not been adequately studied in chimpanzees. Nevertheless, despite the described limitations of the chimpanzee model for HBV infection, these non-human primates have played a major role in the evaluation of the means to prevent and eradicate HBV infection.

Conclusion

Acute HBV infection in chimpanzees, as compared to humans, is characterized by a mild clinical course and occasional delayed resolution. The conventional HBV markers can be identified by immunological probes in blood and liver, as also employed in humans. HBV DNA is always episomal, and integration to the host DNA has not been documented even in long-lasting HBsAg chimpanzee carriers of up to 14 years. Histologically, acute HBV infection is characterized by ballooning degeneration, spotty lobular necrosis and acidophilic bodies, and accumulation of lobular mononuclear cells. Fulminant hepatitis and massive necrosis do not occur. Chronic active hepatitis has never been observed, with one exception in an animal with chronic inflammatory bowel disease. Persistent inflammatory activity in the portal spaces within the limiting plate may however occur, and lasts significantly longer after resolution of hepatoctellular injury. The inability of some chimpanzee HBsAg carriers to clear HBV is poorly characterized and is not associated with cycles of massive necrosis and regeneration observed in humans, a phenomenon that may be linked to the absence of cirrhosis and hepatocellular carcinoma. Despite some significant dissimilarities with human HBV, the chimpanzee model system for HBV has been very useful in the determination of infectivity, characterization of infectious doses, resistance to challenge, effectiveness of passive and active immunization, safety of blood products, immunopathology, clearance of HBV and the molecular characterization of hepatitis δ virus strains.

References

1 Deinhardt F: Hepatitis in primates; in Laufer MA, Bang FA, Maramorosch K, Smith KM (eds): Advance in Virus Research. New York, Academic Press, 1976, vol 20, pp 113–157.
2 Prince AM: Hepatitis-associated antigen: Long-term persistence in chimpanzees; in Goldsmith EI, Moor-Jankowski J (eds): Medical Primatology (Proc 2nd Conf Exp Med Surg Primates, New York 1969). Basel, Karger, 1971, pp 731–739.
3 Barker LF, Chisari FV, McGrath PP, et al: Transmission of viral hepatitis type-B to chimpanzees. J Infect Dis 1973;127:648–662.
4 Blumberg BS, Sutnick AI, London WT: Hepatitis and leukemia: Their relation to Australia antigen. Bull NY Acad Med 1968;44:1566–1586.
5 Hirschman RJ, Shulman NR, Barker LF, Smith KO: Virus-like particles in sera of patients with infectious and serum hepatitis. JAMA 1969;208:1667–1670.
6 Barker LF, Maynard JE, Purcell RH, Hoofnagle JH, Bernquist KR, London WT: Viral hepatitis in experimental animals. Am J Med Sci 1975;270:189–195.
7 Hoofnagle JH, Michalak T, Nowoslawski A, Gerety RJ, Barker LF: Immunofluorescence microscopy in experimentally induced type B hepatitis in the chimpanzee. Gastroenterology 1978;74: 182–187.
8 Popper H, Dienstag JL, Feinstone SM, Alter HJ, Purcell RH: The pathology of viral hepatitis in chimpanzees. Virchow's Arch Pathol Anat Histol 1980;387:91–106.
9 Shikata T, Karasawa T, Abe K: Two distinct types of hepatitis in experimental hepatitis B virus infection. Am J Pathol 1980;99:353–362.
10 Krawczynski K, Prince AM, Nowoslawski A: Immunopathologic aspects of the HBsAg carrier state in chimpanzees. J Med Primatol 1979;8:222–232.
11 Shouval D, Chakraborty PR, Ruiz-Opazo N, Baum S, Spigland I, Muchmore E, Gerber MA, Thung SN, Popper H, Shafritz DA: Chronic hepatitis in chimpanzee carriers of hepatitis B virus: Morphologic, immunologic and viral DNA studies. Proc Natl Acad Sci USA 1980;77:6147–6151.

Daniel Shouval, MD, Director Liver Unit, Hadassah University Hospital,
Jerusalem 91120 (Israel)

Eder G, Kaiser E, King FA (eds): The Role of the Chimpanzee in Research.
Symp, Vienna 1992. Basel, Karger, 1994, pp 180–182

The Role of Chimpanzees in Hepatitis D Research

M. Rizzetto, A. Ponzetto, F. Negro

Department of Gastroenterology, Hospital Molinette, Torino, Italy

The chimpanzee model has been critical in the identification of the hepatitis δ virus (HDV) as a distinct virological entity and to the understanding of the pathogenesis of hepatitis D.

The discovery of HDV originated in the midseventies in Turin, Italy, from the fortuitous observation of the δ antigen (HDAg) in the liver of Italian carriers of the hepatitis B virus (HBV) affected by chronic liver disorders [1]. This finding and the similarities between the pattern of intrahepatic immunohistochemical staining of HDAg and that shown by the HBV core antigen initially suggested that HDAg was an immunological variant of HBV, in keeping with the dozens of antigenic variants of HBV that were described at that time. In view of the inextricable relation with HBV, the story of the δ antigen would have probably ended with the description of an 'Italian' subtype of HBV, were it not for the initiation, at the end of the seventies, of a fortunate collaboration between the Turin group and the National Institutes of Health, Betheseda, Md. As the first step of a collaborative effort, captive chimpanzees were made available for experimental studies on the nature of δ antigen. The chimpanzee was the only animal susceptible to experimental HBV infection and, by inference, to the transmission of HDAg.

In a first battery of normal chimpanzees, inoculation of the serum of an Italian HBV carrier with HDAg in the liver resulted in the simultaneous expression of HDAg and HBV antigens in the liver of the recipient animals [2]. This preliminary experiment, therefore, seemed to imply that HDAg was indeed a component of the HBV virion. However, when in a second series of experiments the same infectious human serum was inoculated into chimpanzees that were already carriers of HBV, HDAg was expressed earlier, to a greater extent and for a longer period of time than in normal animals. In addition, HDAg expression within the liver coincided with a marked decrease in HBV viremia and with the onset of a severe hepatitis, although the animals had been previously asymptomatic in spite of the chronic HBV carriage [2]. This pattern was reminiscent of the biological properties of the defective interfering particles, such as generated in the course of some human viral infections. However, when a unique, small-molecular-weight RNA was isolated from the serum of infected chimpanzees, it became clear that HDAg was associated, rather than with HBV, with a new, defective pathogen different from HBV, yet requiring some obligatory helper function from HBV for in vivo infection [3].

The chimpanzee has continued to play a key informative role throughout all the subsequent progress of our knowledge of the pathobiology of hepatitis D, providing experimental suggestions to the clinical verification as well as experimental confirmation to hypotheses emerging from the clinical and epidemiological scrutiny. Besides confirming the high pathogenic potential of HDV by showing that experimental HDV infections invariably caused an acute hepatitis in the animals and besides elucidating the two mechanisms of coinfection and superinfection, by which HDV is transmitted to normal persons or to HBV carriers, respectively, this animal model has also contributed to the understanding of many pathological, clinical and virological aspects of hepatitis D. It has shown that the severity of liver damage is independent of the infectious dose [4] but correlated with the passage number [5]; in serial transmission studies [5], a decrease in the incubation period was documented together with an increase in pathogenicity. The appearance of disease in correlation with the intrahepatic expression of HDAg has suggested a direct pathogenic effect of HDV on the liver [2]; however, although the histological findings included degenerative lesions of the hepatocytes (hydropic swelling, eosinophilic clumping of the cytoplasms), they consisted also of a conspicuous lymphocytic infiltration of the portal tracts. Characteristic cytoplasmic structures were also noted at the electron microscope level, similarly to what was observed in non-A, non-B hepatitis [6]. Levels of HDV viremia during acute hepatitis D have reached peak values as high as 10^{12} genomes/ml, with a significant decrease in the expression of HBV markers in serum and liver at the time of the florid HDV infection [5]; the higher infectious end points of HDV as compared to HBV further helped to differentiate the two agents. Although in the chimpanzee the acute phase of hepatitis D has lasted up to 7 months, with fluctuating serum alanine aminotransferase levels, and persistent HDV infection has been documented at the colonies of both the National Institutes of Health [7] and of the Centers for Disease Control [8], chronic carriage of the HDV was not accompanied by significant chronic liver damage: in this respect, the animal model has failed so far to reproduce the natural history of chronic hepatitis D in humans. Serum levels of HDV RNA in chronically infected animals are also far lower (approx. 10^7 genomes/ml) than seen during the acute phase of infection [7]. Recent experiments have also raised the possibility that animals which apparently cleared HDV may become reinfected: rechallenge with HDV of chimpanzees previously infected with this virus, which had an acute hepatitis D and became negative for serum HDV RNA, was invariably followed by the reappearance of low levels of viremia [9].

The chimpanzee has also been an invaluable source of material for the molecular characterization of HDV RNA. The first complete sequence of an HDV RNA was obtained from a chimpanzee isolate, taken at the time of an experimental acute hepatitis D [10]. More recently, a chimpanzee was successfully transfected in the liver with a cloned, genomic-length HDV cDNA [11]. This experiment has helped to clarify some aspects of the molecular virology of HDV, in particular the significance of the expression of the two forms of the HDAg.

The chimpanzee animal model is curently providing unique experimental opportunities to understand the relationship between HDV and its pathobiological properties.

References

1 Rizzetto M, et al: Immunofluorescence detection of a new antigen-antibody system (o/anti-o) associated to the hepatitis B virus in the liver and in the serum of HBsAg carriers. Gut 1977; 18:997–1003.

2 Rizzetto M, et al: Transmission of the hepatitis B virus-associated delta antigen to chimpanzees. J Infect Dis 1980;141:590–602.

3 Rizzetto M, et al: Delta antigen: The association of delta antigen with hepatitis B surface antigen and ribonucleic acid in the serum of delta infected chimpanzees. Proc Natl Acad Sci USA 1980; 77:6124–6128.

4 Ponzetto A, et al: Titration of the infectivity of hepatitis D virus in chimpanzees. J Infect Dis 1987; 155:72–78.

5 Ponzetto A, et al: Serial passage of hepatitis delta virus in chronic hepatitis B virus carrier chimpanzees. Hepatology 1988;8:1655–1661.

6 Kamimura T, et al: Cytoplasmic tubular structures in liver of HBsAg carrier chimpanzees infected with delta agent and comparison with cytoplasmic structures in non-A, non-B hepatitis. Hepatology 1983;3:631–637.

7 Negro F, et al: Chronic hepatitis D virus (HDV) infection in hepatitis B virus carrier chimpanzees experimentally superinfected with HDV. J Infect Dis 1988;158:151–159.

8 Fields HA, et al: Experimental transmission of the delta virus to a hepatitis B chronic carrier chimpanzee with the development of persistent delta carriage. Am J Pathol 1986;122:308–314.

9 Negro F, et al: Reappearance of hepatitis delta virus (HDV) replication in chronic hepatitis B virus carrier chimpanzees rechallenged with HDV. J Infect Dis 1989;160:567–571.

10 Wang KS, et al: Structure, sequence and expression of the hepatitis delta (o) viral genome. Nature 1986;323:508–514.

11 Sureau C, et al: Cloned hepatitis delta virus cDNA is infectious in the chimpanzee. J Virol 1989; 63:4294–4297.

Prof. Mario Rizzetto, MD, Department of Gastroenterology, Hospital Molinette,
88 Corso Bramante, I-10126 Torino (Italy)

Eder G, Kaiser E, King FA (eds): The Role of the Chimpanzee in Research.
Symp, Vienna 1992. Basel, Karger, 1994, pp 183–187

Hepatitis C in Chimpanzees and Humans

Harvey J. Alter

Department of Transfusion Medicine, National Institutes of Health,
Bethesda, Md., USA

This chapter will not dwell on the history of the chimpanzee in hepatitis C investigations, but the chimp has been an incredibly important model in this disease. Even before the cloning of the hepatitis C virus (HCV), we knew from chimp experiments that non-A, non-B (NANB) hepatitis was due to a blood-borne transmissible agent; we knew it could be passaged from chimp to chimp; we knew from filtration studies followed by chimp inoculation that the agent was very small; and from Dr. Bradley's assumptions, we even knew that it was related to the togaviruses, a speculation that has proved to be correct. I am not going to detail these critical early studies because I want to discuss some new discoveries, particularly the application of molecular biology to the study of HCV infection in the chimp. To do this, I am going to reverse the usual sequence by beginning with the human and going back to the chimp.

In the typical case of posttransfusion hepatitis, a mild acute hepatitis slowly evolves into varying stages of chronic liver disease. We previously thought, based on alanine aminotransferase (ALT) elevations, that 50% of patients progress to chronic hepatitis, but by measuring HCV RNA, it now appears that 60–80% of patients develop chronic infection. The onset of hepatitis C is generally 6–9 weeks after transfusion. The major ALT elevation occurs in the acute phase and is characteristically followed by a fluctuating ALT pattern as the disease becomes chronic. We have now documented chronic hepatitis persisting over 15 years in many of our patients.

Utilizing second-generation anti-HCV assays, antibody generally appears 5–20 weeks after exposure and then persists, perhaps indefinitely. The average 10-week window until the first appearance of antibody allows some infectious donors to pass the donor screen and delays the specific diagnosis in cases of acute hepatitis. In contrast, HCV RNA is detected very early, usually within 1–2 weeks of exposure. An individual is probably infected and infectious from the point of HCV RNA detection. Viremia and, by implication, infectivity persists indefinitely in the vast majority of patients. Rarely, patients demonstrate complete recovery from HCV infection. In a well-documented case in the NIH series, patient K sustained an acute episode of hepatitis with ALT levels up to 1,500 U/l followed by complete normalization of ALT throughout 14 years of follow-up. In this patient, antibody to c-100 appeared right after the acute phase, persisted at a high level for 5 years and then diminished till no longer detectable 9 years after onset. Antibody to second-generation antigens, c-33 and c-22, however persisted throughout follow-up. HCV RNA was detected by polymerase chain reaction (PCR) 1 week after transfusion and through the preacute and acute phases but then disappeared coincident with biochemical recovery and never recurred after the acute hepatitis. Thus, over a period of years, this patient had detectable anti-HCV antibody in the absence of HCV infection. This is the relatively rare instance of a true recovery from HCV infection. When patient K's acute-phase serum was inoculated into a chimp, the pattern of acute hepatitis with complete recovery and short-lived viremia was again seen. This pattern was reproduced in a second-passage chimpanzee, and we are wondering now whether this particular inoculum is unique in that it uniformly causes a short-duration disease compared to the chronicity induced by most other inocula. In addition, RNA sequencing of this inoculum shows that it is a distinctly different HCV strain than the classic Chiron strain and our Hutchinson strain. It has only a 65% homology with these strains whereas most other strains have 80–90% homology. The relationship of strain to disease outcome needs to be pursued further.

It has been suggested over the years that there might be more than one agent of NANB hepatitis. Experiments by Dr. Yoshisawa, Dr. Bradley, Dr. Tabor and others suggested that the chimp could be reinfected and develop a second episode of NANB hepatitis. The initial feeling was that the second episode was due to a different agent and that there were at least two NANB viruses. We are now able to readdress this issue using molecular biology, particularly to ask the question whether HCV infection induces a protective immune response. These studies were done with Dr. Robert Purcell and Dr. Patricia Farci, and all the PCR and sequencing, the hard part of these studies, was done by Dr. Farci.

We went back to chimpanzee cross-challenge studies that we had performed long ago in collaboration with Dr. Steven Feinstone. Basically, we studied challenges with the same (homologous) or different (heterologous) NANB hepatitis (HCV) inocula.

If one administers the K strain, just as in the human, there is the early onset of PCR activity followed by acute hepatitis and then biochemical recovery followed by the loss of PCR reactivity. Hence, there is a self-limited infection followed by the appropriate antibody responses. In time, the antibody responses abate, PCR becomes negative and the animal appears to have cleared the HCV infection. At this point, we inoculated the F strain. PCR reactivity returned and became persistent, antibodies reappeared and also became persistent; serum enzymes increased only slightly, but there was evidence of new liver damage on biopsy. This pattern suggested either an exacerbation of the initial infection or a new infection. At this point, we could not distinguish these alternatives.

In a second experiment, we challenged with the same HCV strain. Initially, inoculation with the F strain resulted in a self-limited disease; antibody was persistent at low titer but PCR reactivity disappeared. When the animal was challenged with exactly the same inoculum at exactly the same dose, PCR activity returned, enzymes rose slightly and antibody reappeared in high titer. Overall, in the 5 animals challenged, 4 showed what appeared to be two distinct episodes of HCV infection, but the underlying issue was whether the second episode represented new infection or exacerbation of a prior infection.

To distinguish these possibilities, Dr. Farci performed sequencing of the PCR product. RNA was extracted, reverse transcribed and then amplified from the 5′ untranslated region and the hypervariable E1–E2 domain, the latter used for sequencing. In control experiments one could show that the sequence in the human inoculum was identical to that recovered in the infected chimp.

In the challenged animals, when the F strain was inoculated, the F sequence was recovered in the chimp at the time of acute infection. When this animal was challenged with the F strain, sequencing revealed that it was again infected with the homologous strain (F). When we gave the K strain, K sequences were detected in the recipient animal's PCR product. Thus, there did not appear to be an exacerbation of the previous infection but the onset of a new infection. This pattern was reproduced in other animals, even after multiple challenges. Animals that were persistently infected after the initial inoculum did not show new viral sequences on rechallenge.

In summary, these studies demonstrate that in animals that recover from initial infection, rechallenge with a different HCV strain consistently results in the reappearance of viremia; DNA sequencing reveals that the recurrent viremia is not due to reactivation of the original infection but to reinfection with a new HCV strain. In addition, the animal may develop a chronic infection on challenge despite clearance of the initial infection. Thus, one can have an acute infection when first exposed and develop chronic HCV infection on a subsequent exposure.

When the animals were reinfected, antibodies to c-100 returned, but antibodies to envelope did not develop. There does not appear to be a good antibody response to the envelope proteins. Such anti-envelope responses would be the prime candidates for a protective neutralizing antibody. Instead, the antibodies are directed primarily against nonstructural proteins or the core protein. The lack of an adequate immune response to envelope proteins may be a critical determinant in the host's inability to clear HCV infection and hence in the frequent occurrence of persistent infection in both chimps and man.

In all the chimps, reinfection was associated with very mild elevations of ALT; the peak ALT was always considerably lower on rechallenge than during the primary infection. There seems to be protection from a second severe disease episode, even though there is not protection from infection.

I think the implications of this study are considerable. They suggest that HCV does not elicit a protective immune response against reinfection with either the homologous or heterologous strain. This will impinge on vaccine development; such a development will be very difficult both because the virus, as with HIV, is rapidly mutating and because the immune responses are limited and directed against nonprotective epitopes. In a prior vaccine study, performed before the cloning of HCV, we took the H strain inoculum, the only human inoculum with a titer as high as 10^6 CID_{50}/ml, and boiled the serum to inactivate the NANB agent. We then administered the boiled serum in an immunizing schedule (0, 1 and 3 months) combined with alum as adjuvant. The boiled material did not result in hepatitis. Thus, we had inactivated the virus, but when the 'vaccinated' animals were challenged with unheated material, a typical episode of hepatitis occurred. Hence, this three-dose immunization schedule did not protect 2 chimpanzees against a 1-ml intravenous challenge containing 10^6 CID. In a third animal, we concentrated the Hutchinson inoculum twentyfold by pelleting. When this animal was challenged, there was no enzyme elevation and no clinical evidence of infection, but there were cytoplasmic tubular structures observed by electron microscopy. The experiment suggested that immunization attenuated the subsequent infection, but since it involved only 1 chimp, we did not want to make too much of it. Interestingly, Chiron have now used purified HCV antigen as the immunogen and have obtained very similar results in that they have not prevented infection but might have attenuated disease. We need to do PCR and full HCV antibody assays to better delineate the course of infection in our 'immunized' animals.

Overall, the chimpanzee has been an invaluable model for studying NANB and now HCV infection. Indeed, HCV might never have been cloned were it not for chimpanzee experimentation. As in hepatitis B, the chimp will play a vital role in establishing vaccine efficacy, if such a vaccine comes to fruition. If ever a case could be made for the value of animal experimentation, it would be the use of the chimp to unravel and prevent human hepatitis infection. Countless lives have been saved, and not a single chimp has been sacrificed or rendered ill.

Harvey J. Alter, MD, Department of Transfusion Medicine, National Institutes of Health, 9000 Rockville Pike, Building 10, Room 1C711, Bethesda, MD 20892 (USA)

Eder G, Kaiser E, King FA (eds): The Role of the Chimpanzee in Research.
Symp, Vienna 1992. Basel, Karger, 1994, pp 188–191

Identification and Significance of Hepatitis C and E Virus Antigens in Liver Tissue

K. Krawczynski

Experimental Pathology Section, Hepatitis Branch, DVRD/NCID,
Centers for Disease Control, Atlanta, Ga., USA

The successful transmission of hepatitis C and hepatitis E to primates and the subsequent development of suitable experimental models have been critical in the identification of viruses responsible for the majority of non-A, non-B hepatitis worldwide [1]. Extensive immunohistochemical studies have recently led to the morphologic identification of hepatitis C virus (HCVAg) and hepatitis E virus (HEVAg) antigens in the liver. Viral specificity of the antigens in the liver is of significance for clinical and experimental studies on the pathogenesis of blood-borne (hepatitis C) and enterically transmitted (hepatitis E) non-A, non-B hepatitis.

Hepatitis C

HCVAg was detected immunohistochemically using fluorescein-isothiocyanate (FITC) labeled immunoglobulin G fractions from chimpanzee and human sera strongly reactive with recombinant hepatitis C virus (HCV) structural and nonstructural proteins [2]. HCVAg was found in the cytoplasm of individual hepatocytes or groups of liver cells. The antigen had a very fine granular, powder-like pattern with superimposed larger granules of distinct and brilliant fluorescence (fig. 1). The antigen was found in all 9 chimpanzees with acute hepatitis C and in 5 of 10 chimpanzees with chronic HCV infection. HCVAg was observed in 11 of 12 patients with chronic hepatitis C [chronic persistent hepatitis (1/1), chronic active hepatitis (8/9) and active liver cirrhosis (2/2)]. In control studies, HCVAg was not found in liver biopsy specimens obtained from chimpanzees before, shortly after HCV inoculation or during the convalescent phase of disease. HCVAg-negative were liver biopsy specimens

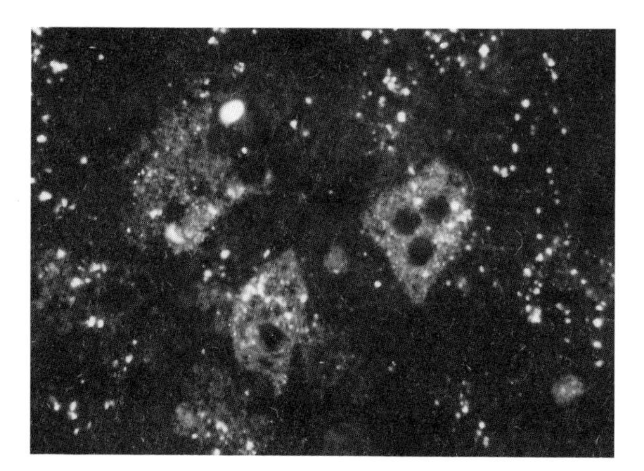

Fig. 1. HCVAg in a frozen section from a liver biopsy specimen obtained from a chimpanzee during the acute phase of experimental infection with HCV US1 strain of the virus. The antigen has been identified by FITC-labeled human polyclonal IgG anti-HCVAg in the cytoplasm of hepatocytes. In addition to HCVAg deposits, coarse granules present in the liver are deposits of lipofuscin with strong autofluorescence.

from chimpanzees (n = 9) and from patients (n = 13) with various etiologies of viral hepatitis: type A, B, E, D and nonviral hepatitis liver conditions. Hepatocellular reactivity of HCVAg was not host derived as evidenced by absorptions of anti-HCVAg FITC-labeled probes with selected host proteins (normal liver homogenate, plasma proteins, red blood cells).

The HCV specificity of the hepatocellular HCVAg and FITC-labeled probes was ascertained by blocking studies with paired serum samples, obtained from 8 infected and uninfected chimpanzees or from 14 patients during the acute and chronic phases of HCV infection. Direct immunomorphologic evidence for the HCV specificity of hepatocellular HCVAg deposits and the FITC-labeled polyclonal anti-HCVAg probe was established in absorption experiments using recombinant HCV structural and nonstructural proteins. HCVAg specificity of the HCV putative NS3 protein, the most prominent component of hepatocellular HCVAg, was documented in absorption experiments with the yeast-expressed HCV core, envelope (E2/NS1), NS3 and NS4 proteins. Absorptions with recombinant HCV nonstructural proteins expressed in *Escherichia coli* (NS3, NS4, NS5) revealed that HCVAg fluorescence in hepatocytes was decreased by NS3 protein.

In chimpanzees infected with a factor-VIII-derived strain of HCV, the virus replication was assessed by identification of HCVAg in hepatocytes and HCV RNA in serum and correlated with IgG and IgM antibody response against HCV structural and nonstructural proteins. In serial samples obtained

Fig. 2. Distinct granular fluorescence of HEVAg in the cytoplasm of hepatocytes detected by immunofluorescence in liver frozen sections of a cynomolgus macaque infected intravenously with the Burma strain of HEV.

during acute hepatitis, HCVAg was observed in liver cells from the 18th to 53rd day after inoculation and correlated with the presence of HCV RNA detected in sequential serum samples. HCV RNA was found up to 14 days before and 2 months after the identification of HCVAg in liver cells. Anti-c33 IgG and transient IgM responses preceded the occurrence of anti-c100 (IgG); anti-HCV-core antibodies were not detected. The features of HCV replication that preceded and overlapped liver pathology were followed by antiviral humoral immune response. In 10 animals with chronic HCV infection, HCVAg (5/10), HCV RNA (6/10) and anti-HCV-core (6/10) were associated with severer features of liver pathology. Anti-c33 (10/10) and anti-c100 (9/10) were regularly observed in chronic HCV infection. The correlation of HCVAg in hepatocytes, HCV RNA in serum, antiviral antibodies and liver pathology enables the evaluation of significance of HCV replication and antigenic expression and virus-specific immune response for the pathogenesis of HCV infection.

Hepatitis E

An antigen (HEVAg) of hepatitis E virus (HEV) was identified in the liver in primates experimentally infected with HEV [3]. HEVAg was observed in the cytoplasm of hepatocytes as granular deposits rarely exceeding 2 μm in diameter (fig. 2). HEVAg was observed in liver biopsy specimens from 5 passages of hepatitis E in cynomolgus macaques, aotus monkeys and from 2 passages of the disease in chimpanzees. Serial liver biopsy specimens from cynomolgus macaques infected intragastrically allowed to evaluate pathogenetic features of HEV infection in experimentally infected primates.

Immunofluorescent probes for identification of HEVAg were prepared from serum previously shown by immune electron microscopy to react with HEV particles. Viral specificity of HEVAg in liver cells was confirmed in a series of absorption experiments and immunologically specific blocking experiments. Preinoculation, acute and convalescent sera from hepatitis E patients and from experimentally infected primates were used in blocking assays. The results of fluorescent antibody blocking and immune electron microscopy assays further suggested that HEVAg is associated with 32- to 34-nm virus-like particles etiologically linked to disease. More recently, recombinant HEV proteins encoded by ORF2 (C2 and C51 proteins), expressed in *E. coli*, were used for absorption procedures to prove viral specificity of HEVAg [Krawczynski et al., in preparation].

The identification of HEVAg facilitated the development of a fluorescent antibody blocking assay for the identification and titration of specific anti-HEVAg antibody [4]. Anti-HEVAg antibody prevalence varied from 70 to 100% in small groups of patients from well-defined and geographically isolated outbreaks. This finding confirms an immunologic cross-reactivity of the tested serum samples and suggests that one virus or class of morphologically and serologically related family of viruses is responsible for hepatitis E in Asia, Africa and North America. Sporadic cases of acute, non-transfusion-transmitted hepatitis E in the USA, Japan and Italy have been found to be serologically unrelated to those derived from well-characterized hepatitis E outbreaks. In these studies only a few acute hepatitis patients, who returned from visits to endemic areas, were found to be serologically related to HEV infection. In the developed countries, the majority of acute non-A, non-B hepatitis is etiologically linked to the new identified HCV or to other hepatotropic viruses such as cytomegalovirus, Epstein-Barr virus and other unidentified viruses.

References

1 Bradley DW, Krawczynski K, Beach MJ, et al: Non-A, non-B hepatitis: Toward the discovery of hepatitis C and E viruses. Semin Liver Dis 1991;11:128–144.

2 Krawczynski K, Beach MJ, Bradley DW, et al: Hepatitis C virus antigen in hepatocytes: Immunomorphologic detection and identification. Gastroenterology 1992;103:622–629.

3 Krawczynski K, Bradley DW: Enterically transmitted non-A, non-B hepatitis: Identification of virus associated antigen in experimentally infected cynomolgus macaques. J Infect Dis 1989;159: 1042–1049.

4 Krawczynski K, Bradley DW, Ajdukiewicz A, et al: Virus associated antigen and antibody of epidemic non-A, non-B hepatitis: Serology of outbreaks and sporadic cases; in Shikata T, Purcell RH, Uchida T (eds): Viral Hepatitis C, D and E. Amsterdam, Elsevier, 1991, pp 229–236.

Kris Krawczynski, MD, PhD, Chief, Experimental Pathology Section, Hepatitis Branch, DVRD/NCID, Centers for Disease Control, Bldg. 6, Room 192, Atlanta, GA 30333 (USA)

Eder G, Kaiser E, King FA (eds): The Role of the Chimpanzee in Research.
Symp, Vienna 1992. Basel, Karger, 1994, pp 192–198

Prospects for Future Research with Chimpanzees

Michale E. Keeling

The University of Texas M.D. Anderson Cancer Center, Bastrop, Tex., USA

The chimpanzees' contributions to biomedical research that benefit human health and well-being have been substantial and are well documented. Although the supply, demand and use of chimpanzees in research fluctuates considerably, medical history confirms that there will always be a need for chimpanzees in biomedical research. The chimpanzees' complete role in diminishing the global impact of our current most significant health threat, the human immunodeficiency virus (HIV), is yet to be determined; but the fact that HIV will replicate in the chimpanzee and is not fatal to the animal would suggest they will play a vital role. The complexity of the animal, and the relationships between production, research, and conservation at times create contradictory objectives. However, by concerted cooperation within the scientific community, I think we can be conditionally optimistic about future use of the chimpanzee in research.

The conditions on which this optimism is based will be presented as a prospectus for the chimpanzee's future role in research. The prospectus will be based upon the assumption that a consortium of diverse, scientific participants can satisfy reasonable expectations that are being mandated by new knowledge about the chimpanzee, technological development and research priorities, regulatory agencies and societal expectations. Specific contributions to create these conditions for continued use will be presented from the perspective of the scientific investigator, the resource and production manager, governmental health agencies, the private health care industry, primate conservationists, zoological societies and the concerned public. The objectives each vested participant must achieve include those over which we have control and those we can only influence.

I think it very fitting that the oldest chimpanzee research institute in the USA (Yerkes Primate Research Center) and the newest and only chimpanzee research institute sponsored by a private company (Immuno AG) play the leadership roles in an effort to revitalize international collaborations in performing responsible biomedical research with the chimpanzee.

I want to approach my conclusions on the prospects for future research with chimpanzees on two fronts:

First, I want to briefly summarize some conclusions I will take from this meeting based upon the programmatic subjects on which we have heard presentations. Perhaps they are similar to some of your conclusions, perhaps in our discussion you can share other conclusions. Next, I will enumerate some optimistic views on future research with the chimpanzee supported by specific contributions several scientific disciplines will have to make as conditions for continued use.

From our first scientific session concerning the ethics, politics, regulations, and comparative similarities of the chimpanzee and the human, I think the following conclusions can be derived:

As brought to our attention by Dr. McCarthy, after years of self-analysis, self-improvement and debate, the appropriateness of performing responsible and humane comparative biomedical research with the chimpanzee continues to be reinforced, even by examples of global consensus dating back to the 1946 Nuremberg Trials. The burden of disproving that the use of animals in research is appropriate is with those who challenge it. In fact, Dr. Cohen described animal research as a *moral obligation* comparative medical scientists have to humans, a prerequisite to responsible human clinical trials. Dr. Cohen's position is that we must protect the humans' right to continued improvements in health and welfare.

As Dr. Horton pointed out, we must guard against complacency concerning the effectiveness of our critics and try to help the public understand the true moderate ground we try to hold. Our research enterprise already answers to 10 or 12 regulatory agencies, and we must be mindful of the impact of additional ambiguous regulations, even if they are performance based.

Although Dr. Hood made the prediction in reference to the human genome project, the point was made that the centers that will play the major future roles in computerized genome sequencing will be those capable of rapid interdisciplinary interaction. I think this will apply equally to future research with the chimpanzee. The centers that will continue to excel in chimpanzee research will be those that are capable of rapid interdisciplinary interaction with the multiple disciplines represented at this symposium and some that are not. We must recognize, accept and nurture the obvious interdependency we have for one another. Our research enterprise also needs to provide the interdisciplinary training referred to in the genome project. Our young scientists must be cross-trained to effectively pursue research in this very complex animal model.

The renewed emphasis on improving the quality of life of captive chimpanzees through psychological enrichment and applied behavioral and reproductive research presents significant opportunities to perform companion research in conjunction with production programs. Several of those programs were reviewed during the session emphasizing the relationship between chimpanzee and human development, behavior and reproduction. These studies must be well designed, objective and capable of being statistically validated.

In one of our discussion sessions, we revisited the issue of concern for monetary commitment to enrichment programs and the elements of a model program. The message is that there is no cookbook, no panacea, no single ideal model. It takes interdisciplinary insight and interaction to develop a customized, easily defended enrichment program that fits your institution's individual needs based upon programmatic goals. A very important point was made that while enrichment programs may seem superfluous to our research goals, they are vitally important in the public's perception of animal research. We must be willing to balance our preference for scientific precision with more subjective elements that improve our public image. The issue of confounding variables associated with psychologically impoverished animals used in sophisticated biomedical research was not discussed, but can affect research quality.

Visions that date back to Dr. R.M. Yerkes in the 1930s and efforts that began in the early 1970s to establish a self-sustaining breeding population of chimpanzees in the USA have reached fruition during the last 2 years. The vision for and support of the National Chimpanzee Breeding and Research Program by the NIH (National Institutes of Health) has resulted in a stable, self-sustaining chimpanzee production colony that is experiencing a 4% annual growth rate and that based upon current projections, can satisfy research demands of the immediate future. The critical role the International Species Inventory System plays in this effort is well recognized.

The emerging role and patterns of chimpanzee research were reviewed and references made to the tremendous potential contributions chimpanzee research can make in the field of infectious diseases (especially HIV and hepatitis), immunology, vaccine development, degenerative diseases and neuroimmunology. Dr. McClure provided an excellent review of the continuing struggle to prevent and treat diseases in the captive chimpanzee population. If I can digress momentarily, I would like to reinforce a new disease concern Dr. McClure referred to briefly. Three cases of spontaneous clinical *leprosy* have been identified recently in the captive population [1]. As men-

- scientific investigators
- governmental health agencies
- private health agencies
- (governmental) regulatory agencies
- resource and production managers
- conservationists
- zoological community
- concerned public (not necessarily critics)

Fig. 1. Consortium of scientists responsible for research use of the chimpanzees.

tioned, PGL-1 serology is being used to survey the populations at risk. There appears to be a disturbing incidence of leprosy exposure in both wild-born and captive-born animals that have been surveyed to date. It appears transmission is occurring within our captive population. This is an insidious disease with latency periods of decades. We have personal experience with effectively treating clinical cases of leprosy, but based upon the human leprosy experience, 5–10 years of double or triple drug therapy are sometimes required. Be cognizant of this new threat to the animal and to your personnel, and please help Dr. Bob Gormus with population surveillance studies if you have the opportunity.

I will not devote much time to presentations concerning viral infection research in the chimpanzee since they are so fresh, but in summary, continued responsible research use of the chimpanzee is vital. The importance of having resources of experimentally naive animals that can tolerate periods of experimental use in a biosafety level 3 setting was identified. We reviewed the overwhelming and convincing characteristics of the chimpanzee model that make it the best assay available for certain diseases, based upon host specific responses to some agents. This enthusiasm for use of the chimpanzee model was tempered with the logic of careful animal model selection, conservation of this vital resource and recognition of the need to provide paybacks for the animals' contributions to science.

Now, if we can address what I referred to in my abstract as my conditional optimism for the future research use of the chimpanzee. The conditions that must be met for continued use of the chimpanzee consist primarily of contributions from the consortium of scientists represented at this symposium and some who are not represented.

Figure 1 represents the diversity of disciplines that must work together to achieve our *dual* responsibility of (1) using the chimpanzee in biomedical research and (2) assuring stable captive populations for the future. We will assume that this diverse community of scientists can interact to satisfy reasonable expectations that are being mandated by new knowledge about the chimpanzee, technological development and research priorities, regulatory agencies and public expectations. Based upon this assumption, I would like to go down our list and present my view of future contributions that will have to be made by each discipline if research use of the chimpanzee is to continue.

Scientific Investigators

Scientists must develop a sensitivity for social pressures and be willing to provide public testimony justifying the need for the chimpanzee as a comparative research model. Scientists must consider the numerous options for improving the quality of life for experimental animals when designing experimental protocols. They must develop insights into enrichment and training options and be willing to coordinate research activities closely with the animal colony manager.

Governmental Health Agencies

These agencies must continue to support and strengthen long-range plans to fund critical resource programs like the National Chimpanzee Breeding and Research Program of the National Center for Research Resources. They must nurture international efforts with European and Japanese governments in protecting and responsibly managing the captive and, indirectly, the wild chimpanzee populations.

Regulatory Agencies

Regulatory agencies (AAALAC, USDA, OPRR, NIH, IAMC) should promote efforts to standardize space and performance requirements specifically for the chimpanzee, not just for all nonhuman primates, by weight range. Regulatory agencies should avoid additional ambiguous regulations. Instead we should organize consensus conferences and committees to develop specific definitions and criteria for promoting psychological well-being in the chimpanzee. In cooperation with major chimpanzee holders, regulatory agencies should sponsor education and training retreats for site visitors and inspectors who are routinely visiting facilities that maintain chimpanzees.

Private Health Agencies

Agencies should become more interactive with chimpanzee production and holding facilities to establish fair market value for chimpanzees and share in the costs of long-range management by awarding endowments or purchasing chimpanzee shares. Private health agencies should identify common goals and interest of collaborating governmental or university groups and educate them to the need for and mechanisms of protecting proprietary information. They should assist inexperienced collaborating agencies in identifying and soliciting private funding.

Resource and Production Facilities

Chimpanzee holders must identify capital improvement funds to build new facilities to cost-effectively maintain the growing number of HIV experimental veterans and to avoid the logjams in our limited biohazard containment facilities. Chimpanzee holders and producers must be more aggressive about changing the public and other scientific disciplines' perception of the US National Chimpanzee Breeding and Research Program. Producers must develop detailed 20-year production and use plans and explore more creative approaches to facility design, stable long-range funding, management, care and use. Those capable of conducting animal model research must identify agencies or institutions that will fund studies of chimpanzee reproductive biology, including embryo cryopreservation and reversible cost-effective contraception. All chimpanzee producers must establish specific guidelines for maintaining genetic diversity in a self-sustaining population during high and low experimental demand periods. Realistic long-range projections of research needs must be developed to avoid production lag times and undesirable shortages of research animals. Chimpanzee producers should collaborate in developing a standardized range of value (fair market value) for domestically produced chimpanzees based upon age, sex, experimental history, type of research proposed, ownership transfer and need for long-term maintenance or endowment. A standardized guide for valuation of chimpanzees would certainly assist investigators and funding agencies in long-range planning and budgeting and would enhance collaborations between public and private user institutions and holding and producing facilities.

Zoological Community

Although not represented at this conference, the zoological community can make contributions to responsible use of the chimpanzee in biomedical research. The zoological community should continue to strengthen the species survival program they have developed for the chimpanzee. They can significantly assist educating the public to the long-range benefits the chimpanzee will realize if zoos collaborate with research facilities that have production programs. The zoological community represents an existing collaborative bridge to the conservation community.

Conservation Community

Again, although not represented at this conference, the conservation community should play a role in the experimental use of the chimpanzee. Conservationists should become more interactive with other scientific disciplines that are sincerely dedicated to conservation goals. Our combined strength will enhance our dual responsibility of conservation in the wild and experimental use of the captive population to the benefit of man. The NIH-sponsored chimpanzee captive breeding program will make significant contributions to the chimpanzee's conservation in the wild.

Concerned Public (Not Necessarily Critics)

The consortium of scientists dealing with the chimpanzee must nurture the public majority that is willing to support good biomedical research with the chimpanzee. This is best achieved by intensive promotional and educational programs that are delivered in plain vanilla language and includes programs for elementary school children. We need to work harder to build collaborative bridges with groups we normally consider adversarial. A unified front on major points of agreement could generate significant funding for those efforts.

If the world is to continue to benefit from chimpanzee research, a strong international, interdisciplinary cooperative with common goals must be established. We must abandon our tendency to develop chimpanzee research plans in 5-year segments, similar to our research programs. We must recognize the need to develop 20-year chimpanzee use plans and explore more creative approaches to funding facility design, management, care and use, as well as public perception.

Reference

1 Gormus BJ, Keyu X, Alford PL, Lee DR, Hubbard GB, Eichberg JW, Meyers WM: A serologic study of naturally acquired leprosy in chimpanzees. Int J Lepr 1991;59:450–457.

Dr. Michale E. Keeling, UTMDACC Science Park, Department of Veterinary Resources Route 2, Box 151-B1, Bastrop, TX 78602 (USA)

Subject Index